Gas Turbine Engineering

Gas Turbine Engineering

Rosario Douglas

NY Research Press
New York

Published by NY Research Press
118-35 Queens Blvd., Suite 400,
Forest Hills, NY 11375, USA
www.nyresearchpress.com

Gas Turbine Engineering
Rosario Douglas

International Standard Book Number: 978-1-63238-809-4 (Hardback)

Cataloging-in-Publication Data

Gas turbine engineering / Rosario Douglas.
p. cm.
Includes bibliographical references and index.
ISBN 978-1-63238-809-4
1. Gas-turbines. 2. Jet engines. 3. Turbines. 4. Engineering. I. Douglas, Rosario.
TJ778 .G37 2020
621.433--dc23

Printed in China.

TABLE OF CONTENTS

PREFACE

This book has been written, keeping in view that students want more practical information. Thus, my aim has been to make it as comprehensive as possible for the readers. I would like to extend my thanks to my family and co-workers for their knowledge, support and encouragement all along.

Gas turbine is a type of continuous combustion, internal combustion engine. The main elements that are common to all gas turbine engines are an upstream rotating gas compressor, a combustor and a downstream turbine on the same shaft as the compressor. Some of the most common types of gas turbines are jet engines, turboprop engines, aeroderivative gas turbines and industrial gas turbines. Gas turbine engineering deals with the design, selection, maintenance and operation of gas turbines. It studies the underlying principles of gas turbine operations as well as the economic considerations and implications of operating these machines. The topics included in this book on gas turbine engineering are of utmost significance and bound to provide incredible insights to readers. It is an upcoming field of science that has undergone rapid development over the past few decades. Those in search of information to further their knowledge will be greatly assisted by this book.

A brief description of the chapters is provided below for further understanding:

Chapter - Introduction

Gas turbine refers to the combustion engine that converts natural gas to mechanical energy and produces electric current. Some of its major components are compressor, combustion chamber, turbine, etc. This is an introductory chapter which will briefly introduce all the significant aspects of gas turbines as well as its principles.

Chapter - Types of Gas Turbines

There are different types of gas turbines which depend on the aircraft engine overhaul. It includes turboprop engines, aeroderivative gas turbines, turboshaft engines, radial gas turbines, microturbines, etc. The topics elaborated in this chapter will help in gaining a better perspective about these types of gas turbines.

Chapter - Engine Control and Noise Reduction

Gas turbine engine control is one of the main components of gas turbine as it assures operational safety and reliability in aircraft engines. Some of the noise reduction techniques in gas turbines are scarf inlets, forward swept fans, soft stator vanes, chevron nozzles, etc. This chapter has been carefully written to provide an easy understanding of the gas turbine engine control and noise reduction techniques.

Chapter - Gas Power Cycles

Gas power cycles include heat transfer, work transfer, pressure variations, temperature variations, volume variations and entropy variations. There are various gas power cycles namely Brayton cycle, Rankine cycle, Carnot cycle, Diesel cycle, etc. All these different gas power cycles have been carefully analyzed in this chapter.

Chapter - Applications

There are a wide-range of applications of gas turbines in electric power generation in industries as well as in marine, locomotive, and automotive propulsion. This chapter closely examines these applications of gas turbines to provide an extensive understanding of the subject.

Rosario Douglas

Introduction

CHAPTER 1

Gas turbine refers to the combustion engine that converts natural gas to mechanical energy and produces electric current. Some of its major components are compressor, combustion chamber, turbine, etc. This is an introductory chapter which will briefly introduce all the significant aspects of gas turbines as well as its principles.

GAS TURBINES

A gas turbine is a type of turbine that uses pressurized gas to spin it in order to generate electricity or provide kinetic energy to an airplane or jet. The process to do so is called the Brayton cycle. In all modern gas turbines, the pressurized gas is created by the burning of a fuel like natural gas, kerosene, propane or jet fuel. The heat generated by this fuel expands air which flows through the turbine to supply useful energy.

A jet engine gas turbine. The flow is from left to right, with labeling of parts in figure.

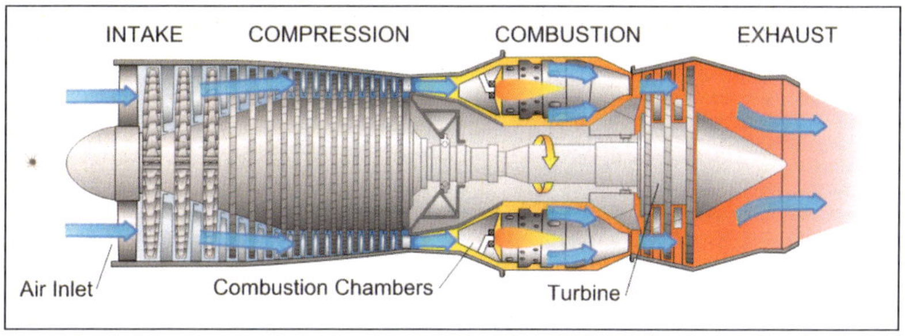

A diagram of a gas turbine engine.

Gas turbines are theoretically simple, and have three main parts as seen in figure:

- Compressor- Takes in air from outside of the turbine and increases its pressure.
- Combustor- Burns the fuel and produces high pressure and high velocity gas.
- Turbine- Extracts the energy from the gas coming from the combustor.

Compressor

In figure, air is sucked in from the left and input to the compressor which consists of many rows of fan blades. In some turbines, the pressure of the air can increase by a factor of 30.

Combustor

The high pressure air flows into this area, which is where the fuel is introduced. The fuel gets injected constantly into this part in order for the energy through the turbine to be constant.

Turbine

The turbine is connected to the compressor blades by a shaft, and they spin separately. The compressor connects to the turbine which is connected to an output shaft, and because the turbine spins separately, it can get up to tremendous speeds due to the hot gas flowing through it. This final shaft generates enormous amounts of horsepower, with large airplane turbines generating nearly 110000 hp - twice the power generated by the Titanic.

SOME BASIC PRINCIPLES

As air passes through a gas turbine engine, aerodynamic and energy requirements demand changes in the air's velocity and pressure. During compression, a rise in the air pressure is required, but not an increase in its velocity. After compression and combustion have heated the air, an increase in the velocity of gases is necessary in order for the turbine rotors to develop power. The size and shape of the ducts through which the air flows affect these various changes. Where a conversion from velocity to pressure is required, the passages are divergent. Conversely, if a conversion from pressure to velocity is needed, a convergent duct is used.

Before further discussion, an explanation of convergent ducts, divergent ducts, and the behavior of air within these ducts should be made. An understanding of the difference between static pressure (P_s), impact pressure, (P_i), and total pressure (P_t) is also needed.

The difference between static, impact, and total pressures is as follows. Static pressure is the force per unit area exerted on the walls of a container by a stationary fluid. An example is the air pressure within a car tire. Impact pressure, on the other hand, is the force per unit area exerted by fluids in motion. Impact pressure is a function of the velocity of the fluid. An example of impact pressure is the pressure exerted on one's hand held outside a moving car's window. Total pressure is the sum of static and impact pressures.

Figure illustrates the methods used to measure pressures. Part (a) illustrates the measurement of static pressure. Static pressure will not take into account the velocity of the air. Part (b) illustrates the measurement of total pressure, which accounts for both static pressure and the pressure due to the moving fluid (impact pressure). In order to obtain impact pressure, the value of the static pressure is subtracted from the value of total pressure.

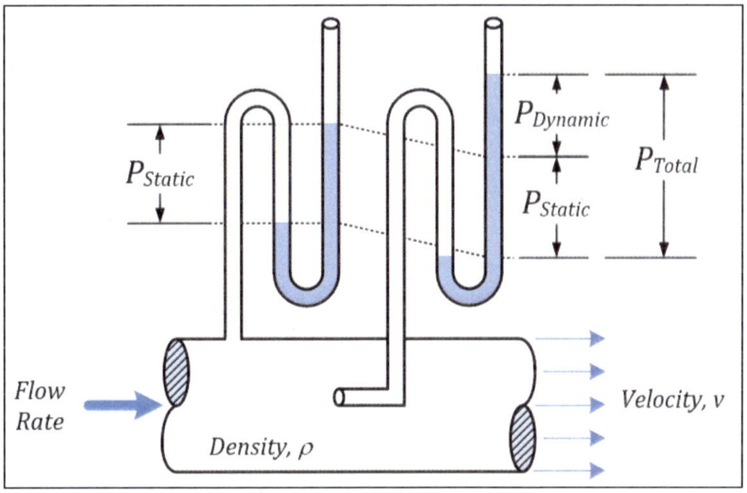

Pressure Management technique.

Figure shows the principle of divergent ducts, where energy is neither being added or taken away, but where the gaseous energy is being converted from velocity to pressure and temperature. There is a velocity decrease as air flows from a small inlet to a larger outlet. As velocity decreases, impact pressure (P_i) also decreases. Since no energy is added or subtracted from the system, total pressure (P_t) for the air remains constant and static pressure (P_s) increases. One way of viewing this is that the impact pressure is converted to static pressure; thus, a static pressure rise is seen as air flows through a divergent duct and is compressed. A temperature rise is also noticed since compression is a heating process.

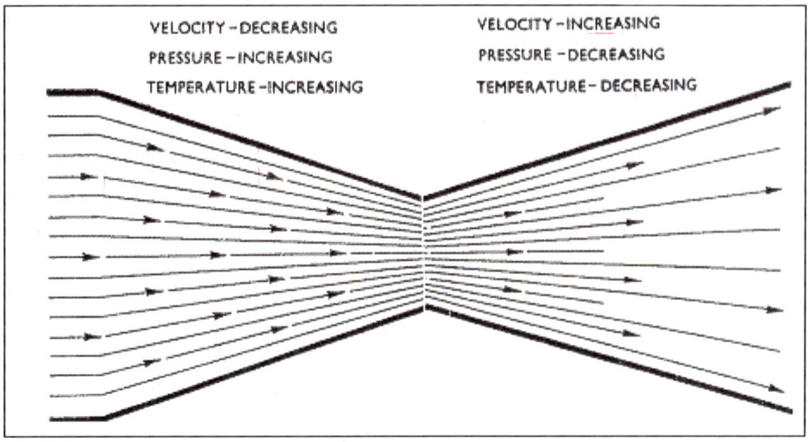

Divergent duct and Convergent Duct.

The convergent duct operates exactly in reverse of the divergent duct. Figure shows the principle of convergent ducts, where energy is neither being added or taken away, but where the gaseous energy is being converted from pressure and temperature to velocity. There is a velocity increase as air flows from a large inlet to a smaller outlet. As velocity increases, impact pressure also increases. Since no energy is added or subtracted from the system, total pressure remains constant and static pressure decreases. One way of viewing this is that the static pressure is converted to impact pressure; thus, a static pressure decrease is seen as air flows through a convergent duct and goes through expansion. A temperature drop is associated with any expansion process.

Even though the static and impact pressures are changing as fluids flow through either convergent or divergent ducts, the total pressure does not change. This is true if fluid friction is neglected and energy is not added or taken away from the fluid flow. In actuality, there will be a slight decrease in total pressure because of fluid frictional losses.

COMPONENT OF GAS TURBINE ENGINES

Early gas turbines employed centrifugal compressors, which are relatively simple and inexpensive. They are, however, limited to low pressure ratios and cannot match the efficiencies of modern axial-flow compressors. Accordingly, centrifugal compressors are used today primarily in small industrial units.

An axial-flow compressor is the reverse of a reaction turbine. The blade passages, which look like twisted, highly curved airfoils, must exert a tangential force on the fluid with the pressures on one side of the blade higher than on the other. For subsonic flow, an increase in pressure requires the flow area to also increase, thus reducing the flow velocity between the blade passages and diffusing the flow. A row of compressor blades must be viewed as a set of closely spaced, highly curved airfoil shapes with which airflow strongly interacts. There will not only be a rise in pressure along the blades but a variation between them as well. Flow friction, leakage, wakes produced by the previous sets of blades, and secondary circulation or swirl flows all contribute to losses in a real unit. Tests of stationary blade assemblies, known as cascades, can be performed in special wind tunnels, but actual blade arrangements in a rotating assembly require special test setups or rigs.

Blades must be designed not only to have the correct aerodynamic shape but also to be light and not prone to critical vibrations. Recent advances in compressor (and turbine) blade design have been aided by extensive computer programs.

While moderately large expansion-pressure ratios can be achieved in a reaction-turbine stage, only relatively small pressure increases can be handled by a compressor stage typically pressure ratios per stage of 1.35 or 1.4 to 1 in a modern design. Thus, compressors require more stages than turbines. If higher stage pressure ratios are attempted, the flow will tend to separate from the blades, leading to turbulence, reduced pressure rise, and a "stalling" of the compressor with a concurrent loss of engine power. Unfortunately, compressors are most efficient close to this so-called surge condition, where small disturbances can disrupt operation. It remains a major challenge to the designer to maintain high efficiency without stalling the compressor.

As the air is compressed, its volume decreases. Thus the annular passage area should also decrease if the through-flow velocity is to be kept nearly constant i.e., the blades have to become shorter at higher pressures. An optimum balance of blade-tip speeds and airflow velocities often requires that the rotational speed of the front, low-pressure end of the compressor be less than that of the high-pressure end. This is achieved in large aircraft gas turbines by "spooled" shafts where the shaft for the low-pressure end, driven by the low-pressure portion of the turbine, is running at a different speed within the hollow high-pressure compressor/turbine shaft, with each shaft having its own bearings. Both twin- and triple-spool engines have been developed.

Combustion Chamber

Air leaving the compressor must first be slowed down and then split into two streams. The smaller stream is fed centrally into a region where atomized fuel is injected and burned with a flame held in place by a turbulence-generating obstruction. The larger, cooler stream is then fed into the chamber through holes along a "combustion liner" (a sort of shell) to reduce the overall temperature to a level suitable for the turbine inlet. Combustion can be carried out in a series of nearly cylindrical elements spaced around the circumference of the engine called cans, or in a single annular passage with fuel-injection nozzles at various circumferential positions. The difficulty of achieving nearly uniform exit-temperature distributions in a short aircraft combustion chamber can be alleviated in stationary applications by longer chambers with partial internal reversed flow.

Turbine

The turbine is normally based on the reaction principle with the hot gases expanding through up to eight stages using one- or two-spooled turbines. In a turbine driving an external load, part of the expansion frequently takes place in a high-pressure turbine that drives only the compressor while the remaining expansion takes place in a separate, free turbine connected to the load.

High-performance aircraft engines usually employ multiple spools. A recent large aircraft-engine design operating with an overall pressure ratio of 30.5:1 uses two high-pressure turbine stages to drive 11 high-pressure compressor stages on the outer spool, rotating at 9,860 revolutions per minute, while four low-pressure turbine stages drive the fan for the bypass air as well as four additional low-pressure compressor stages through the inner spool turning at 3,600 revolutions per minute. For stationary units, a total of three to five total turbine stages is more typical.

High temperatures at the turbine inlet and high centrifugal blade stresses necessitate the use of special metallic alloys for the turbine blades. Such alloys are sometimes grown as single crystals. Blades subject to very high temperatures also must be cooled by colder air drawn directly from the compressor and fed through internal passages. Two processes are currently used: (1) jet impingement on the inside of hollow blades, and (2) bleeding of air through tiny holes to form a cooling blanket over the outside of the blades.

Control and Start-up

In a gas-turbine engine driving an electric generator, the speed must be kept constant regardless of the electrical load. A decrease in load from the design maximum can be matched by burning less fuel while keeping the engine speed constant. Fuel flow reduction will lower the exit temperature of the combustion chamber and, with it, the enthalpy drop available to the turbine. Although this reduces the turbine efficiency slightly, it does not affect the compressor, which still handles the same amount of air. The foregoing method of control is substantially different from that of a steam turbine, where the mass flow rate has to be changed to match varying loads.

An aircraft gas-turbine engine is more difficult to control. The required thrust, and with it engine speed, may have to be changed as altitude and aircraft speed are altered. Higher altitudes lead to lower air-inlet temperatures and pressures and reduce the mass flow rate through the engine. Aircraft now use complex computer-driven controls to adjust engine speed and fuel flow while all critical conditions are monitored continuously.

For start-up, gas turbines require an external motor which may be either electric or, for stationary applications, a small diesel engine.

Other Design Considerations

Many other aspects enter into the design of a modern gas-turbine engine, of which only a few examples can be given. Much attention must be paid, especially in a multispool unit, to the design of all bearings, including the thrust bearings that absorb axial forces, and to the lubrication system. As an engine is started up and becomes hot, components elongate or "grow," thereby affecting passage clearances and seals. Other considerations include bleeding air from the compressor and ducting it for turbine-blade cooling or for driving accessories.

BASIC OPERATING PRINCIPLES

Gas turbine engines derive their power from burning fuel in a combustion chamber and using the fast-flowing combustion gases to drive a turbine in much the same way as the high-pressure steam drives a steam turbine. A simple gas turbine is comprised of three main sections: a compressor, a combustor, and a power turbine. The gas turbine operates on the principle of the Brayton cycle, where compressed air is mixed with fuel and burned under constant pressure conditions. The resulting hot gas is allowed to expand through a turbine to perform work.

Introduction

As the principle of the gas turbine, a working gas (air) is compressed by a compressor and heated by combustion energy of the fuel at the first. The working gas becomes the high temperature and high pressure. The engine converts the energy of working gas into the rotating energy of the blades, making use of the interaction between the gas and the blades.

As shown in the below figure, there are two types of gas turbine. One is the open cycle type (internal type) and another is the closed cycle type (external type). Basic components of both types are the air compressor, a combustor, and the turbine.

The gas turbine can handle a larger gas flow than that of the reciprocating internal combustion engines, because it utilizes a continued combustion. Then the gas turbine is suitable as the high power engine. The gas turbine for airplanes (called a jet engine) makes use of this advantage.

The gas turbine operates on the principle of the Brayton cycle and one variation of this basic cycle is the addition of a regenerator. A gas turbine with a regenerator (heat exchanger) recaptures some of the energy in the exhaust gas, preheating the air entering the combustor. This cycle is typically used on low-pressure ratio turbines, and the resulting hot gas is allowed to expand through a turbine to perform work. In a 33% efficient gas turbine, almost 2/3 of this work is spent compressing the air; the rest is available for other work such as mechanical drive or electrical generation. Figure is schematic of such principle.

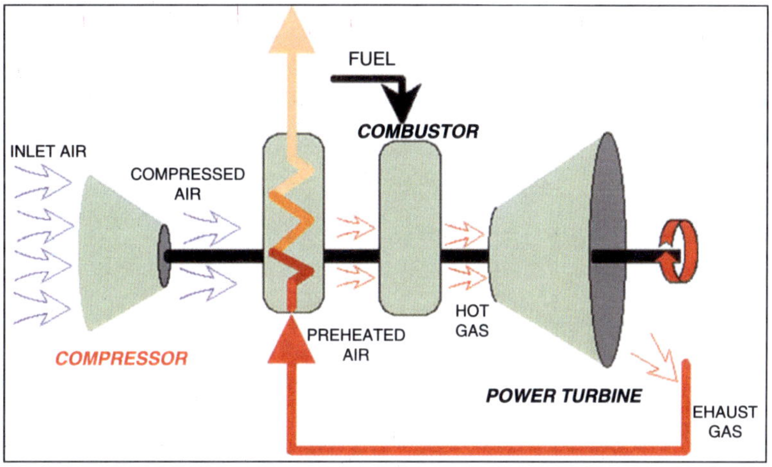

Schematic of solar centaur/3500 horsepower class.

Gas turbines with high-pressure ratios can use an intercooler to cool the air between stages of compression, allowing you to burn more fuel and generate more power. Remember, the limiting factor on fuel input is the temperature of the hot gas created, because of the metallurgy of the first stage nozzle and turbine blades. With the advances in materials technology, this physical limit is always climbing. Figures below are illustrations of a gas turbine, using an intercooling (heat exchanger), while figures is presenting a gas turbine with reheater.

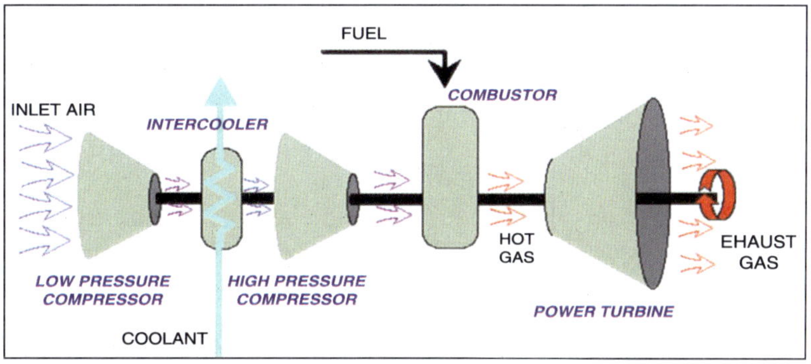

Illustration of a simple gas turbine with intercooling and combustor.

The gas turbine can handle a larger gas flow than that of the reciprocating internal combustion engines, because it utilizes a continued combustion. Then the gas turbine is suitable as the high power engine. The gas turbine for airplanes (called a jet engine) makes use of this advantage.

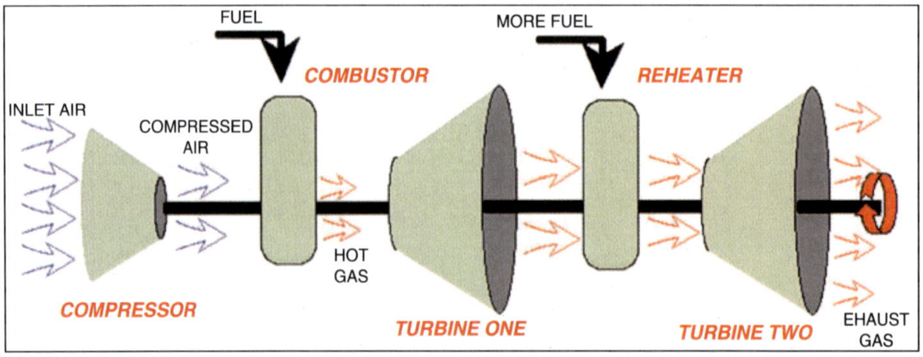

Schematic of gas turbine with reheater.

Generally speaking gas turbine is divided into two categories as follows:

- Open cycle gas turbine.
- Closed cycle gas turbine.

Both of these two cycles are presented in figure.

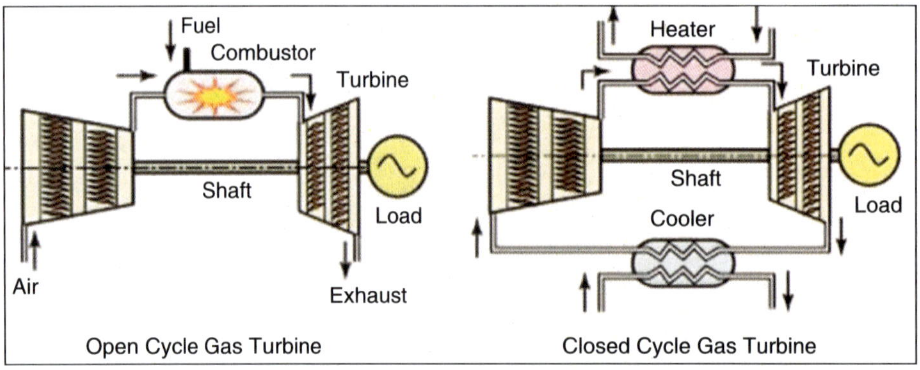

Configuration of open and close cycle gas turbine.

In case of jet engine power plant, as we said, they drive their power from burning fuel in a combustion chamber and using the fast-flowing combustion gases to drive a turbine in much the same way as the high-pressure steam drives a steam turbine.

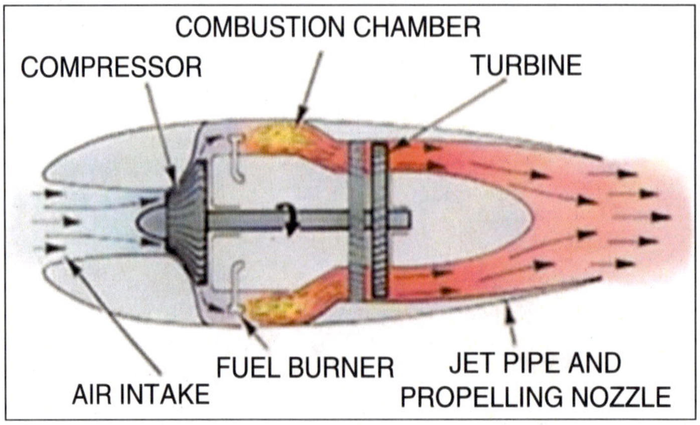

A gas turbine power plant.

One major difference however is that the gas turbine has a second turbine acting as an air compressor mounted on the same shaft. The air turbine (compressor) draws in air, compresses it, and feeds it at high pressure into the combustion chamber increasing the intensity of the burning flame.

It is a positive feedback mechanism. As the gas turbine speeds up, it also causes the compressor to speed up forcing more air through the combustion chamber which in turn increases the burn rate of the fuel sending more high-pressure hot gases into the gas turbine increasing its speed even more. Uncontrolled runaway is prevented by controls on the fuel supply line which limit the amount of fuel fed to the turbine thus limiting its speed.

The thermodynamic process used by the gas turbine is known as the Brayton cycle. Analogous to the Carnot cycle in which the efficiency is maximized by increasing the temperature difference of the working fluid between the input and output of the machine, the Brayton cycle efficiency is maximized by increasing the pressure difference across the turbine. The gas turbine is comprised of three main components: a compressor, a combustor, and a turbine. The working fluid, air, is compressed in the compressor (adiabatic compression—no heat gain or loss) and then mixed with fuel and burned by the combustor under relatively constant pressure conditions in the combustion chamber (constant pressure heat addition). The resulting hot gas expands through the turbine to perform work (adiabatic expansion). Much of the power produced in the turbine is used to run the compressor, and the rest is available to run auxiliary equipment and do useful work. The system is an open system because the air is not reused so that the fourth step in the cycle, cooling the working fluid, is omitted.

Gas turbines have a very high power to weight ratio and are lighter and smaller than internal combustion engines of the same power. Though they are mechanically simpler than reciprocating engines, their characteristics of high-speed and hightemperature operation require high-precision components and exotic materials making them more expensive to manufacture.

One big advantage of gas turbines is their fuel flexibility. They can be adapted to use almost any flammable gas or light distillate petroleum products such as gasoline (petrol), diesel, and kerosene (paraffin) which happen to be available locally, though natural gas is the most commonly used fuel. Crude and other heavy oils can also be used to fuel gas turbines if they are first heated to reduce their viscosity to a level suitable for burning in the turbine combustion chambers.

Gas turbines can be used for large-scale power generation. Examples are applications delivering 600 MW or more from a 400 MW gas turbine coupled to a 200 MW steam turbine in a cogenerating installation. Such installations are not normally used for baseload electricity generation, but for bringing power to remote sites such as oil and gas fields. They do however find use in the major electricity grids in peak shaving applications to provide emergency peak power.

Low-power gas turbine generating sets with capacities up to 5 MW can be accommodated in transportation containers to provide mobile emergency electricity supplies which can be delivered by truck to the point of need.

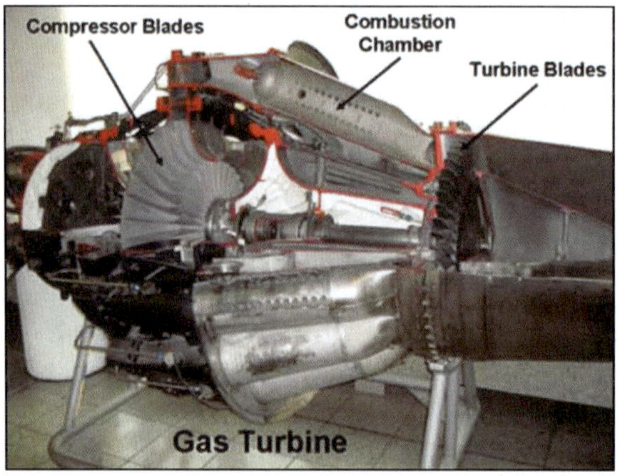

Gas turbine aero engine.

Combined Cycle Power Conversion for New Generation Reactor Systems

A number of technologies are being investigated for the Next Generation Nuclear Plant that will produce heated fluids at significantly higher temperatures than current-generation power plants. The higher temperatures offer the opportunity to significantly improve the thermodynamic efficiency of the energy conversion cycle. One of the concepts currently under study is the molten salt reactor. The coolant from the molten salt reactor may be available at temperatures as high as 800–1000 °C. At these temperatures, an open Brayton cycle combined with a Rankine bottoming cycle appears to have some strong advantages. Thermodynamic efficiencies approaching 50% appear possible. Requirements for circulating cooling water will be significantly reduced. However, to realistically estimate the efficiencies achievable, it is essential to have good models for the heat exchangers involved as well as the appropriate turbo machinery. This study has concentrated on modeling all power conversion equipment from the fluid exiting the reactor to the energy releases to the environment.

Combined cycle power plants are currently commercially available. General Electric STAG™ (steam turbine and generator) systems have demonstrated high thermal efficiency, high reliability/availability, and economic power generation for application in baseload cyclic duty utility service.

Heat recovery-type steam and gas turbine combined cycle systems are the economic choice for gas or oil-fired power generation.

Incorporation with environmentally clean gasification system is extending their economic application to low-cost solid fuel utilization. The features contributing to their outstanding generation economics are:

High thermal efficiency	High reliability
Low installed cost	High availability
Fuel flexibility—wide range of gas and liquid fuels	Short installation time
Low operation and maintenance cost	High efficiency in small capacity increments
Operating flexibility—base, midrange, daily start	

In electricity generating applications, the turbine is used to drive a synchronous generator which provides the electrical power output but because the turbine normally operates at very high rotational speeds of 12,000 r.p.m. or more, it must be connected to the generator through a high ratio reduction gear since the generators run at speeds of 1000 r.p.m. or 1200 r.p.m. depending on the alternating current (AC) frequency of the electricity grid. Gas turbine power generators are used in two basic configurations:

- Simple Systems: This system consists of the gas turbine driving an electrical power generator. The following figure depicts such a configuration.
- Combined Cycle Systems: These systems are designed for maximum efficiency in which the hot exhaust gases from the gas turbine are used to raise steam to power a steam turbine with both turbines being connected to electricity generators.

In both cases as part of turbine performance and as turbine power output, we need to minimize the size and weight of the turbine for a given output power, and the output per pound of airflow should be maximized. This is obtained by maximizing the air flow through the turbine which in turn depends on maximizing the compressor pressure ratio. The main factor governing this is the pressure ratio across the compressor which can be as high as 40:1 in modern gas turbines. In simple cycle applications, pressure ratio increases translate into efficiency gains at a given firing temperature, but there is a limit since increasing the pressure ratio means that more energy will be consumed by the compressor.

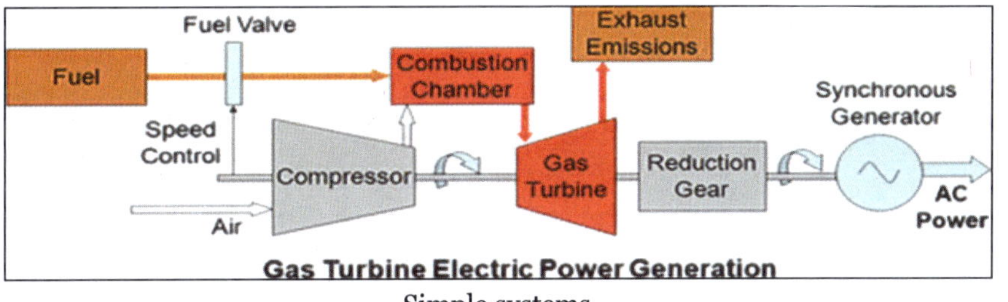

Simple systems.

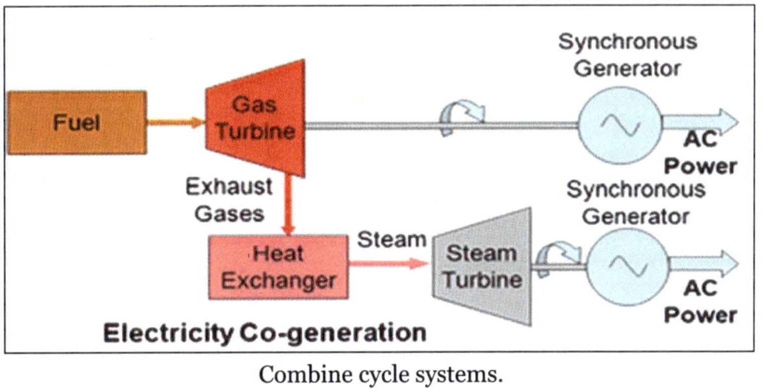

Combine cycle systems.

System Efficiency and Turbine Cycles

Thermal efficiency is important because it directly affects the fuel consumption and operating costs:

- Simple Cycle Turbines.

A gas turbine consumes considerable amounts of power just to drive its compressor. As with all cyclic heat engines, a higher maximum working temperature means greater efficiency (Carnot's Law), but in a turbine it also means that more energy is lost as waste heat through the hot exhaust gases whose temperatures are typically well over 500 °C. Consequently, simple cycle turbine efficiencies are quite low. For a heavy plant, design efficiencies range between 30% and 40%. (The efficiencies of aero engines are in the range 38–42% while low-power micro-turbines (<100 kW) achieve only 18–22%.) Although increasing the firing temperature increases the output power at a given pressure ratio, there is also a sacrifice of efficiency due to the increase in losses due to the cooling air required to maintain the turbine components at reasonable working temperatures.

Generation 3 GE Combined Cycle systems

60Hz STAG Combined Cycle Experience With "F" Technology Gas Turbine

Table: Third generation combined cycle experience.

Country	Installation	Configuration	COD	Output (MW)
USA	Virginia Power #7	S107F	1990	214
USA	Virginia Power #8	S107F	1992	218
Korea	KEPCO Seo-Inchon#1 & #2	8xS107F	1992	1887
USA	Sithe Independence	2xS207FA	1995	1062
USA	Tampa Electric, polk Co.	S107FA	1996	313
Korea	KEPCO Seo-Inchon #3 & #4	2xS207FA	1996	1004
USA	US Gen. Co., Hermiston	2xS107FA	1996	425
USA	Crockett Cogen	S107FA*	1996	202/248
Mexico	CFE Samalayuca	3xS107FA*	1998	506
USA	Cogentrix, Clark Co.	S107FA*	1998	254
USA	Ft. St. Vrain	S207FA	1999	487
Korea	KEPCO, POSCO	S207FA	1999	498
Columbia	EPM LaSierra	S207FA	2001	478
USA	Bucksport Energy	S107FA	2001	176
USA	Westbrook	S207FA	2001	528
USA	Santee Cooper	S207FA	2001	600
Korea	Pusan	S207FA	2003/4	2000

*Single-Shaft Combined Cycle.
Number Of gas turbines = 45 Units.
Installed Capacity = 12,411 MW.
"Combined Cycle Development and Future", David L. Chase, GER-4206, 2001.

increasing the firing temperature increases the output power at a given pressure ratio, there is also a sacrifice of efficiency due to the increase in losses due to the cooling air required to maintain the turbine components at reasonable working temperatures:

- Combined cycle turbines.

It is however possible to recover energy from the waste heat of simple cycle systems by using the exhaust gases in a combined cycle system to heat steam to drive a steam turbine electricity generating set. In such cases the exhaust temperature may be reduced to as low as 140 °C enabling efficiencies of up to 60% to be achieved in combined cycle systems. In combined cycle applications, pressure ratio increases have a less pronounced effect on the efficiency since most of the improvement comes from increases in the Carnot thermal efficiency resulting from increases in the firing temperature.

Thus simple cycle efficiency is achieved with high-pressure ratios. Combined cycle efficiency is obtained with more modest pressure ratios and greater firing temperatures.

Modeling the Brayton Cycle

Any external combustion or heat engine system is always at a disadvantage to an internal combustion system. The internal combustion systems used in current jet engine and gas turbine power systems can operate at very high temperatures in the fluid and cool the structures containing the fluid to achieve high thermodynamic efficiencies. In an external energy generation system, like a reactor powered one, all of the components from the core to the heat exchangers heating the working fluid must operate at a higher temperature than the fluid. This severely limits the peak cycle temperature compared to an internal combustion system. This liability can be overcome to a certain extent by using multiple expansion turbines and designing highly efficient heat exchangers to heat the working fluid between expansion processes similar to reheaters in steam systems.

Typically the combustion chamber in a gas turbine involves a pressure drop of 3–5% of the total pressure. Efficient liquid salt to air heat exchangers can theoretically be designed with a pressure drop of less than 1%. This allows three to five expansion cycles to achieve a pressure drop comparable to a combustion system. Multiple turbines operating at different pressures have been common in steam power plants for a number of years. In this study one to five gas turbines operating on a common shaft were considered.

Multiple expansion turbines allow a larger fraction of the heat input to be provided near the peak temperature of the cycle. The exhaust from the last turbine is provided to the Heat Recovery Steam Generator (HRSG) to produce the steam used in the Rankine bottoming cycle. The hot air after it passes through the HRSG is exhausted to the atmosphere. A detailed comparison of this system was made with a recuperated stand-alone Brayton cycle, and the dual cycle appears to be more efficient for open systems.

Modeling the Rankine Cycle

The Rankine cycle was modeled with the standard set of components including the HSRG, a steam turbine, condenser, and high-pressure pump. Multiple reheat processes were considered. There is a slight efficiency advantage to include two reheat processes as per a fairly standard practice in today's power plants.

The major limitation on the size of the steam system is the enthalpy available from high-temperature air above the pinch point where the high-pressure water working fluid starts to vaporize. Below this point, there is still a significant enthalpy in the air which is readily available to heat the high-pressure water. There does not appear to be an advantage to including feed water heaters in the cycle to bring the high-pressure water up to the saturation point. The possibility that an intercooler could be inserted between the two stages of a split compressor was considered. The cooling fluid for the intercooler was the high-pressure water coming out of the water pump.

This process would combine the function of the traditional intercooler with the preheating of a typical feed water heater. The effect of this addition to the two cycles had a marginal effect on the overall system efficiency and likely is not worth the cost, or effort, to implement.

The Combined Brayton-Rankine Cycle

The combined cycle unit combines the Rankine (steam turbine) and Brayton (gas turbine) thermodynamic cycles by using heat recovery boilers to capture the energy in the gas turbine exhaust gases for steam production to supply a steam turbine as shown in the Fig."combined cycle cogeneration unit." Process steam can be also provided for industrial purposes.

Fossil fuel-fired (central) power plants use either steam or combustion turbines to provide the mechanical power to electrical generators. Pressurized hightemperature steam or gas expands through various stages of a turbine, transferring energy to the rotating turbine blades. The turbine is mechanically coupled to a generator, which produces electricity.

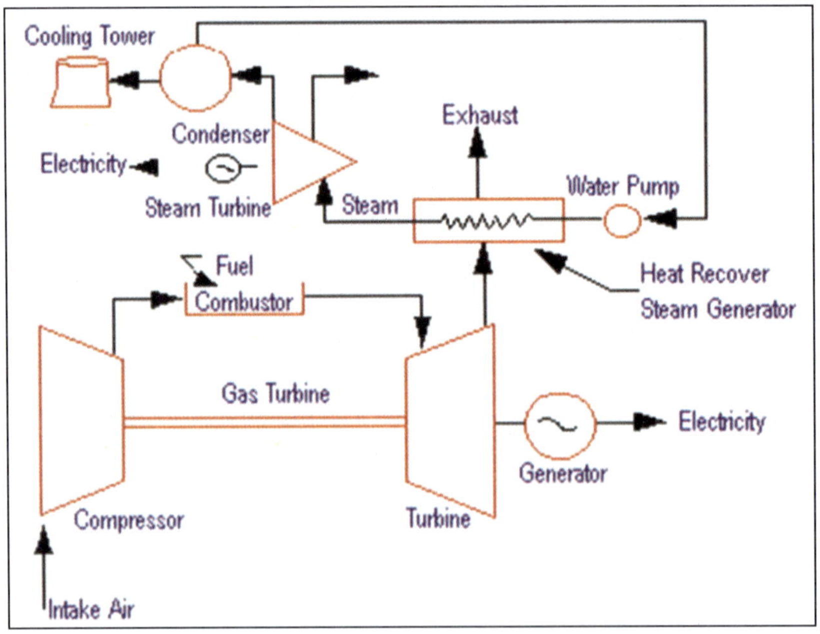

The combined cycle Brayton and Rankine cycle cogeneration unit.

The Brayton cycle efficiency is quite low primarily because a substantial amount of the energy input is exhausted to surroundings. This exhausted energy is usually at a relatively high temperature, and thus it can be used effectively to produce power. One possible application is the combined Brayton Rankine cycle in which the hightemperature exhaust gases exiting the gas turbine are used to supply energy to the boiler of the Rankine cycle, as illustrated in figure Note that the temperature T_9 of the Brayton cycle gases exiting the boiler is less than the temperature T_3 of the Rankine cycle steam exiting the boiler; this is possible in the counterflow heat exchanger, the boiler.

Single and Multi-shaft Design

The gas turbine can be designed in a single or multi-shaft configuration. In the single-shaft case, the gas turbine is designed with roughly equal pressure ratios across all expansion stages which are mechanically coupled to the gas compressor and generator and operate at the generator speed (normally 3600 or 1800 rpm for 60 Hz electrical systems and 3000 or 1500 rpm for 50 Hz electrical systems). In a multi-shaft configuration, the compressor is mechanically driven by a set of expansion stages sized to produce the amount of mechanical work required by the compressor, so that this shaft is not connected to the electrical generator and can rotate at different speeds. The air produced from this gas-generator is heated and directed to a turbo-generator: a final expansion stage on a separate shaft that rotates at the optimal generator speed. Combined cycle gas turbine (CCGT) power plant suppliers configure turbine generators in a number of different arrangements.

Multi-shaft and single-shaft configurations allow customization to optimize plant performance, capital investment, construction and maintenance access, operating convenience, and minimum space requirements.

The development of large F-class gas turbines during the past decade went hand in hand with manufacturers' efforts to standardize combined cycle power plant (CCPP) configurations, striving to best use the new technology. The single-shaft power train (SSPT) arrangement was first conceived for applications using gas turbines over 250 megawatts. Only later the concept was extended to smaller units in the range of 60 megawatts. The new SSPT arrangement allowed single blocks of up to 450 megawatts to be built. SSPTs contributed the most to the power plants aiming at cost savings and project time reductions and thus at lower risk. In SSPT arrangements, the gas turbine and the steam turbine are coupled to a common generator on a single shaft, whereas in multi-shaft power train blocks (MSPT) up to three gas turbines and their allocated boilers and generators share a common steam turbine. SSPT and MSPT are both built for 50 and 60 Hz markets. The main benefits of the new concept highlighted by manufacturers are higher operation flexibility, smaller footprint, simplified control, shorter run-up time, more standardized peripheral systems, and higher efficiency and availability. This development requires that, in addition to new technical issues related to the gas turbine design, insurers look at a great number of aspects when covering entire combined cycle power plants.

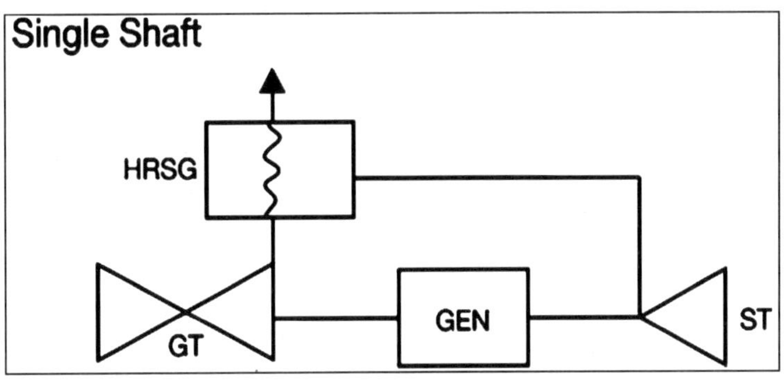

Combined cycle single-shaft arrangements.

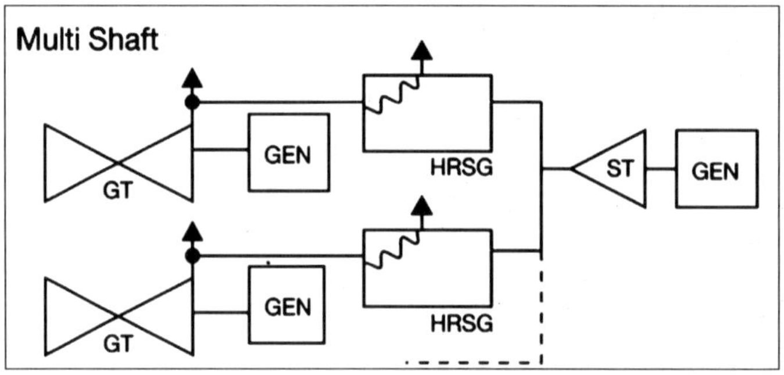

Combined cycle multi shaft arrangements.

In a multi-shaft combined cycle plant, there are generally several gas turbines with HRSGs generating steam for a single steam turbine. The steam and gas turbines use separate shafts, generators, set-up transformers, and so on. By combining the steam production of all the HRSGs, a large steam volume enters the steam turbine, which generally raises the steam turbine efficiency.

Modern gas turbines achieve higher output with higher exhaust temperatures. With the large gas turbines on the market, one steam turbine per gas turbine or one steam turbine for two gas turbines is common.

If one steam turbine per gas turbine is installed, the single-shaft application is the most common solution gas turbine and steam turbine driving the same generator.

A plant with two gas turbines can be built either with two gas turbines on one steam turbine configuration (multi-shaft) or as a plant with two gas turbines, each in a single-shaft configuration. In either scenario usage of clutch as a synchronous self-shifting device has a significant impact on components used in the respective combined cycle power plant (CCPP).

A combined cycle single-shaft configuration with the generator between the two turbines enables installation of a clutch between steam turbine and generator. This means the clutch engages in that moment when the steam turbine speed tries to overrun the rigidly couple gas turbine generator and disengages if the torque transmitted from the steam turbine to generator becomes zero.

The clutch allows startup and operation of the gas turbine without driving the steam turbine. This results in a lower starting power requirement and eliminates certain safety measures for the steam turbine (e.g., cooling steam or sealing steam).

Furthermore, clutch implementation provides design opportunities for accommodating axial thermal expansion. The clutch itself compensates for a portion of axial displacement, and the two thrust bearings allow selective distribution of the remaining axial expansion (reducing tip clearance losses). In addition, it allows more operational flexibility such as gas turbine simple cycle operation or early preventive maintenance (PM) activities on gas turbine during steam turbine cool down.

Generations of combined cycle power plant equipment that are manufactured by General Electric are also divided into two basic configurations:

- Single shaft.
- Multiple shaft.

The single-shaft combined cycle system consists of one gas turbine, one steam turbine, one generator, and one Heat Recovery Steam Generator (HRSG), with the gas turbine and steam turbine coupled to a single generator in a tandem arrangement.

Single shaft arrangements where a gas turbine and a steam turbine drive a single generator are often preferred because they offer a more compact plant at a lower cost. Many include an SSS (name of manufacture in United Kingdom) clutch to disconnect the steam turbine and to allow the gas turbine/generator to be operated separately.

Advantages of this arrangement are as follows:

- Simple startup:
  - Standard gas turbine (GT) start.
  - Reduced time to generation.
  - No cooling steam required.
  - Reduced starting power.
  - Reduced emissions.
  - Standardized design.
  - Simplified torsion analysis.

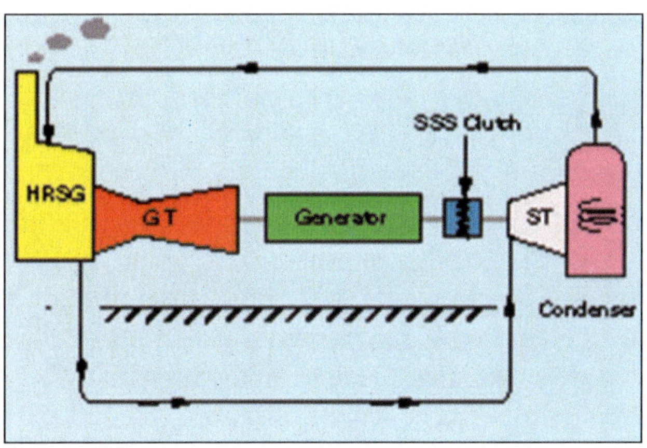

Single shaft CCGT arrangement using an SSS clutch.

- Increased flexibility:
  - Simplified commissioning.
  - Steam turbine trips do not stop power generation.
  - Maintenance of gas turbine (GT) is possible during steam generation (SG) cooling.
  - No cooling steam required.
  - Reduced starting power.
  - Reduced emissions.

  - Standardized design.
  - Simplified torsion analysis.
- Optimized shutdown.
  - Shutdown steam turbine (ST) at reduced gas turbine (GT) power.

Multi-shaft combined cycle systems have one or more gas turbine generators and HRSGs that supply steam through a common heater to a separate single steam turbine generator unit. Both configurations perform their specific functions, but the single-shaft configuration excels in the baseload and midrange power generation applications.

The multi-shaft combined cycle system configuration is most frequently applied in phased installations in which the gas turbines are installed and operated prior to the steam cycle installation and where it is desired to operate the gas turbines independent of the steam system. The multi-shaft configuration was applied most widely in the early history of heat recovery combined cycles primarily because it was the least departure from the familiar conventional steam power plants. The single-shaft combined cycle system has emerged as the preferred configuration for single phase applications in which the gas turbine and steam turbine installation and commercial operation are concurrent.

Multi-shaft systems have one or more gas turbine-generators and HRSGs that supply steam through a common header to a separate single steam turbine-generator. In terms of overall investment, a multi-shaft system is about 5% higher in costs.

The primary disadvantage of multiple-stage combined cycle power plant is the number of steam turbines, condensers and condensate systems, and perhaps the cooling towers and circulating water systems required by the bottoming cycle.

A Heat Recovery Steam Generator (HRSG) is a heat exchanger or series of heat exchangers that recovers heat from a hot gas stream and uses that heat to produce steam for driving steam turbines or as process steam in industrial facilities or as steam for district heating.

An HRSG is an important part of a combined cycle power plant (CCPP) or a cogeneration power plant. In both of those types of power plants, the HRSG uses the hot flue gas at approximately 500–650 °C from a gas turbine to produce highpressure steam. The steam produced by an HRSG in a gas turbine combined cycle power plant is used solely for generating electrical power. However, the steam produced by an HRSG in a cogeneration power plant is used partially for generating electrical power and partially for district heating or for process steam.

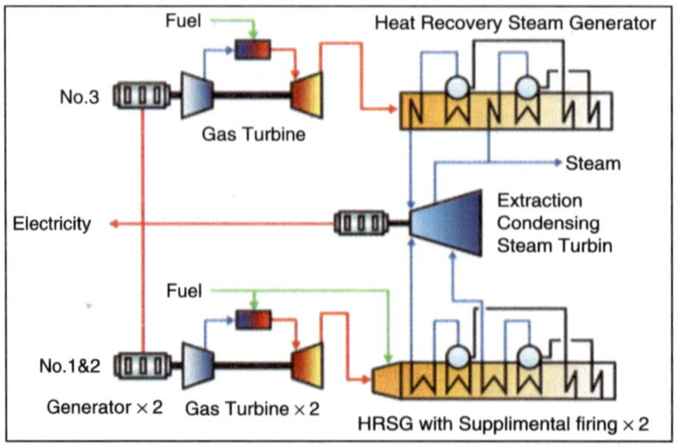

Combined cycle power plant with multi shaft configuration.

The combined cycle power plant, schematically depicted in figure is so named because it combines the Brayton cycle for the gas turbine and the Rankine cycle for the steam turbines. About 60% of the overall electrical power generated in a CCPP is produced by an electrical generator driven by the gas turbine and about 40% is produced by another electrical generator driven by the high-pressure and low-pressure steam turbines. For large-scale power plants, a typical CCPP might use sets consisting of a gas turbine driving a 400 MW electricity generator and steam turbines driving a 200 MW generator (for a total of 600 MW), and the power plant might have two or more such sets.

The primary components of the heat exchangers in an HRSG are the economizer, the evaporator and its associated steam drum, and the superheater as shown in figure An HRSG may be in horizontal ducting with the hot gas flowing horizontally across vertical tubes as in figure or it may be in vertical ducting with the hot gas flowing vertically across horizontal tubes. In either horizontal or vertical HRSGs, there may be a single evaporator and steam drum or there may be two or three evaporators and steam drums producing steam at two or three different pressures. Figure depicts an HRSG using two evaporators and steam drums to produce high-pressure steam and low-pressure steam, with each evaporator and steam drum having an associated economizer and superheater. In some cases, supplementary fuel firing may be provided in an additional section at the front end of the HRSG to provide additional heat and higher temperature gas.

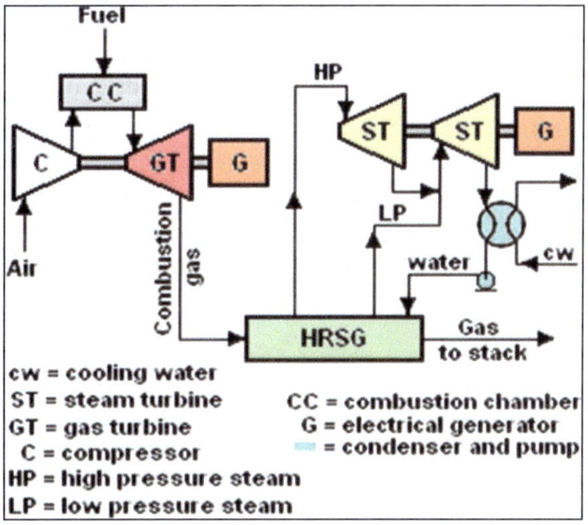

HRSG for multi shaft combined cycle power plant.

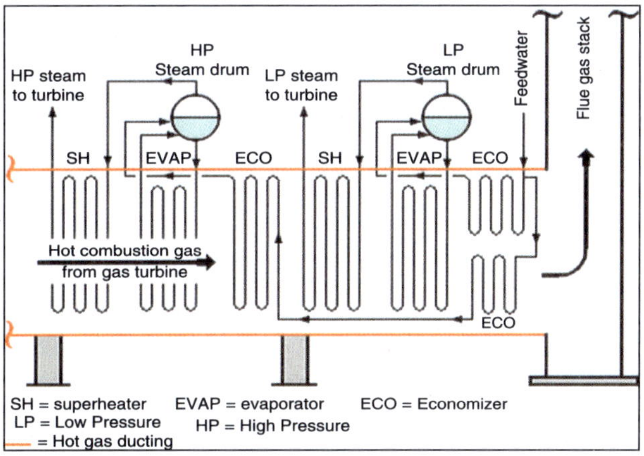

Heat Recovery Steam Generator (HRSG).

There are a number of other HRSG applications. For example, some gas turbines are designed to burn liquid fuels (rather than gas) such as petroleum naphtha or diesel oil, and others burn the syngas (synthetic gas) produced by coal gasification in an integrated gasification combined cycle plant commonly referred to as an IGCC plant. As another example, a combined cycle power plant may use a diesel engine rather than a gas turbine. In almost all such other applications, HSRGs are used to produce steam to be used for power generation.

Working Principle of Combined Cycle Gas Turbine (CCGT)

The first step is the same as the simple cycle gas turbine plant. An open cycle gas turbine has a compressor, a combustor, and a turbine. For this type of cycle, the input temperature to the turbine is very high. The output temperature of the flue gases is also very high. This is high enough to provide heat for a second cycle which uses steam as the working medium, i.e., thermal power station.

Air Inlet

This air is drawn though the large air inlet section where it is cleaned, cooled, and controlled. Heavy-duty gas turbines are able to operate successfully in a wide variety of climates and environments due to inlet air filtration systems that are specifically designed to suit the plant location.

Under normal conditions, the inlet system has the capability to process the air by removing contaminants to levels below those that are harmful to the compressor and turbine.

In general the incoming air has various contaminants. They are:

In gaseous state, contaminants are:

- Ammonia,
- Chlorine,
- Hydrocarbon gases,
- Sulfur in the form of H_2S, SO_2,
- Discharge from oil cooler vents.

In liquid state contaminants are:

- Chloride salts dissolved in water (sodium, potassium),
- Nitrates,
- Sulfate,
- Hydrocarbons.

In solid state contaminants are:

- Sand, alumina, and silica,

- Rust,
- Road dust, alumina, and silica,
- Calcium sulfate,
- Ammonia compounds from fertilizer and animal feed operations,
- Vegetation, airborne seeds.

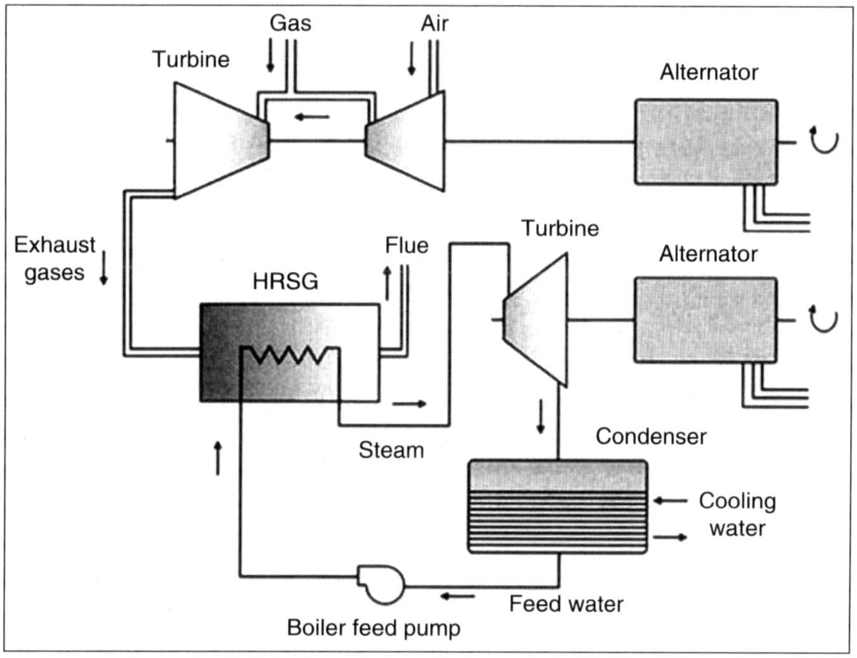

Working principle of combined cycle gas turbine (CCGT) plant.

Corrosive agents:

- Chlorides, nitrates, and sulfates can deposit on compressor blades and may result in stress corrosion attack and cause corrosion pitting. Sodium and potassium are alkali metals that can combine with sulfur to form a highly corrosive agent and that will attack portions of the hot gas path. The contaminants are removed by passing through various types of filters which are present on the way.
- Gas phase contaminants such as ammonia or sulfur cannot be removed by filtration. Special methods are involved for this purpose.

Turbine Cycle

The air is purified then compressed and mixed with natural gas and ignited, which causes it to expand. The pressure created from the expansion spins the turbine blades, which are attached to a shaft and a generator, creating electricity.

In the second step, the heat of the gas turbine's exhaust is used to generate steam by passing it through a Heat Recovery Steam Generator (HRSG) with a live steam temperature between 420 °C and 580 °C.

Heat Recovery Steam Generator

In the Heat Recovery Steam Generator, highly purified water flows in tubes, and the hot gases pass around them to produce steam. The steam then rotates the steam turbine and coupled generator to produce electricity. The hot gases leave the HRSG at around 140 centigrade and are discharged into the atmosphere.

The steam condensing and water pump systems are the same as in the steam power plant.

Typical Size and Configuration of CCGT

The combined cycle system includes single-shaft and multi-shaft configurations. The single-shaft system consists of one gas turbine, one steam turbine, one generator, and one Heat Recovery Steam Generator (HRSG), with the gas turbine and steam turbine coupled to the single generator on a single shaft.

Multi-shaft systems have one or more gas turbine-generators and HRSGs that supply steam through a common header to a separate single steam turbine generator. In terms of overall investment, a multi-shaft system is about 5% higher in costs.

Efficiency of CCGT Plant

Roughly the steam turbine cycle produces one third of the power and gas turbine cycle produces two thirds of the power output of the CCPP. By combining both gas and steam cycles, high input temperatures and low output temperatures can be achieved. The efficiency of the cycles adds, because they are powered by the same fuel source.

To increase the power system efficiency, it is necessary to optimize the HRSG, which serves as the critical link between the gas turbine cycle and the steam turbine cycle with the objective of increasing the steam turbine output. HRSG performance has a large impact on the overall performance of the combined cycle power plant.

The electric efficiency of a combined cycle power station may be as high as 58% when operating new and at continuous output which are ideal conditions. As with single cycle thermal units, combined cycle units may also deliver low temperature heat energy for industrial processes, district heating, and other uses. This is called cogeneration and such power plants are often referred to as a combined heat and power (CHP) plant.

The efficiency of CCPT is increased by supplementary firing and blade cooling. Supplementary firing is arranged at HRSG, and in gas turbine a part of the compressed air flow bypasses and is used to cool the turbine blades. It is necessary to use part of the exhaust energy through gas to gas recuperation. Recuperation can further increase the plant efficiency, especially when gas turbine is operated under partial load.

Fuels for CCPT Plants

The turbines used in combined cycle plants are commonly fueled with natural gas, and it is more versatile than coal or oil and can be used in 90% of energy applications. Combined cycle plants are usually powered by natural gas, although fuel oil, synthesis gas, or other fuels can be used.

Emission Control

Selective catalytic reduction (SCR):

- To control the emissions in the exhaust gas so that it remains within permitted levels as it enters the atmosphere, the exhaust gas passes through two catalysts located in the HRSG.
- One catalyst controls carbon monoxide (CO) emissions and the other catalyst controls oxides of nitrogen (NOx) emissions. Aqueous ammonia, in addition to the SCR (a mixture of 22% ammonia and 78% water), is injected into system to even further reduce levels of NOx.

Advantages of Combined Cycle Power Plants are:

Fuel Efficiency

In conventional power plants, turbines have a fuel conversion efficiency of 33% which means two thirds of the fuel is burned to drive the turbine off. The turbines in combined cycle power plant have a fuel conversion efficiency of 50% or more, which means they burn about half amount of fuel as a conventional plant to generate same amount of electricity.

Low Capital Costs

The capital cost for building a combined cycle unit is two thirds the capital cost of a comparable coal plant.

Commercial Availability

Combined cycle units are commercially available from suppliers anywhere in the world. They are easily manufactured, shipped, and transported.

Abundant Fuel Source

The turbines used in combined cycle plants are fueled with natural gas, which is more versatile than a coal or oil and can be used in 90% of energy publications. To meet the energy demand nowadays, plants are not only using natural gas but also using other alternatives like biogas derived from agriculture.

Reduced Emission and Fuel Consumption

Combined cycle plants use less fuel per kWh and produce fewer emissions than conventional thermal power plants, thereby reducing the environmental damage caused by electricity production. Comparable with coal-fired power plant, burning of natural gas in CCPT is much cleaner.

Potential Applications in Developing Countries

The potential for combined cycle plant is with industries that require electricity and heat or stem.

Disadvantages of combined cycle power plants are:

- The gas turbine can only use natural gas or high-grade oils like diesel fuel.

- Because of this the combined cycle can be operated only in locations where these fuels are available and cost effective.

Gas Turbine Technology and Thermodynamics

The turbine entry temperature, T_c, is fixed by materials technology and cost. (If the temperature is too high, the blades fail.) figures Below show the progression of the turbine entry temperatures in aero engines. Figure is from Rolls-Royce and figure is from Pratt & Whitney. Note the relation between the gas temperature coming into the turbine blades and the blade melting temperature.

For a given level of turbine technology (in other words given maximum temperature), a design question is:

- What should the compressor temperature ratio across compressor between point a and b (TR) be?
- TR = Tb/Ta.
- What criterion should be used to decide this?
- Maximum thermal efficiency?
- Maximum work?

We examine this issue below using figure as baseline guide.

The problem is posed in Fig., which shows two Brayton cycles. For maximum efficiency, we would like TR as high as possible. This means that the compressor exit temperature approaches the turbine entry temperature. The network will be less than the heat received; as $T_b \rightarrow T_c$ the heat received approaches zero and so does the network.

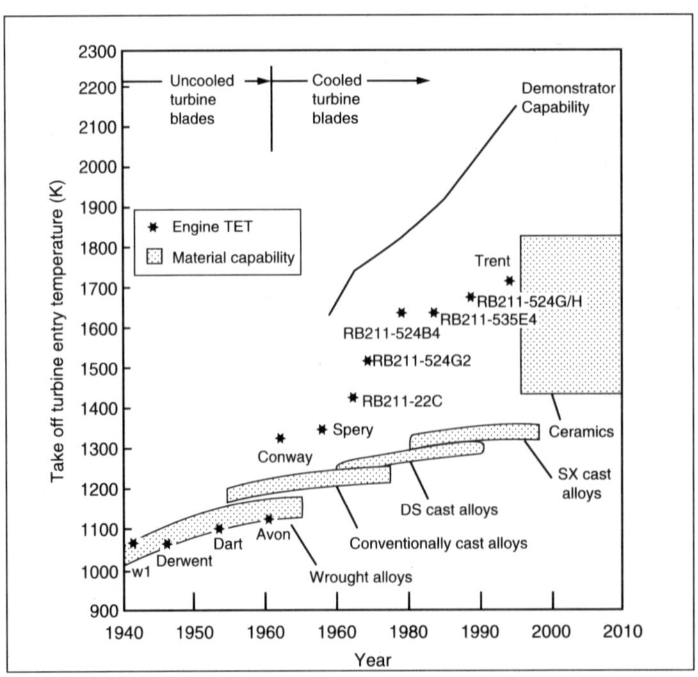

Rolls-Royce high-temperature technology.

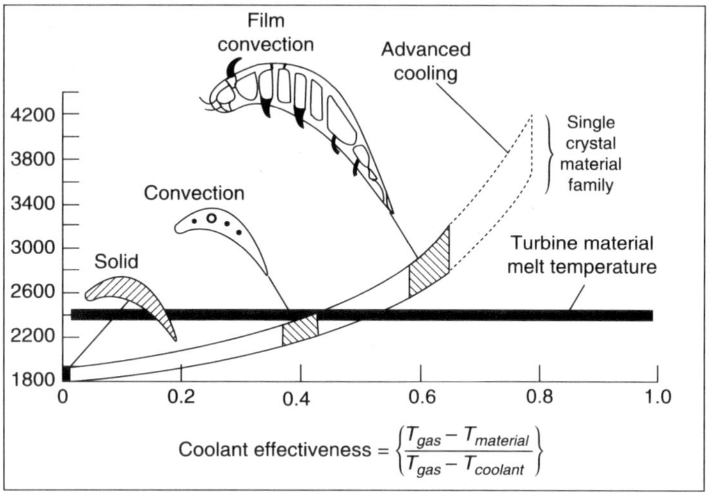

Turbine blade cooling technology.

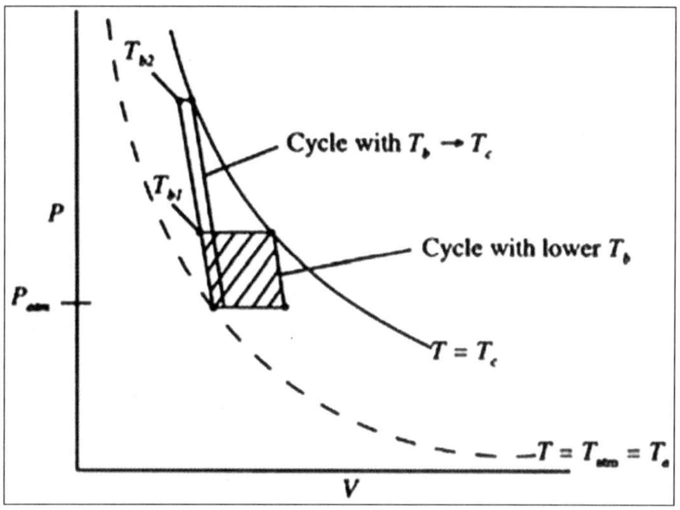

Efficiency and work of two Brayton cycle engines.

The network in the cycle can also be expressed as ∫ Pdv, evaluated in traversing the cycle. This is the area enclosed by the curves, which is seen to approach zero as $T_b \to T_c$.

The conclusion from either of these arguments is that a cycle designed for maximum thermal efficiency is not very useful in that the work (power) we get out of it is zero.

A more useful criterion is that of maximum work per unit mass (maximum power per unit mass flow). This leads to compact propulsion devices. The work per unit mass is given by:

$$\text{Work/ Unit Mass } = c_p\left[(T_c - T_b) - (T_b - T_a)\right]$$

where T_c is the maximum turbine inlet temperature (a design constraint) and T_a is atmospheric temperature. The design variable is the compressor exit temperature, T_b, and to find the maximum as this is varied, we differentiate the expression for work with respect to T_b:

$$\frac{d\,\text{Work}}{dT_b} = c_p\left[\frac{dT_c}{dT_b} - 1 - \frac{dT_d}{dT_b} + \frac{dT_a}{dT_b}\right]$$

The first and the fourth terms on the right-hand side of the above equation are both zero (the turbine entry temperature is fixed, as is the atmospheric temperature). The maximum work occurs where the derivative of work with respect to T_b is zero and it can be seen as:

$$\frac{d\,\text{Work}}{dT_b} = 0 = -1 - \frac{dT_d}{dT_b}$$

To use Eq. $\frac{d\,\text{Work}}{dT_b} = 0 = -1 - \frac{dT_d}{dT_b}$ we need to relate Td and Tb. But we know that:

$$\frac{T_d}{T_a} = \frac{T_c}{T_b} \text{ or } T_d = \frac{T_a T_c}{T_b}$$

Hence, we can write:

$$\frac{dT_d}{dT_b} = -\frac{T_a T_c}{T_b^2}$$

Plugging Eq. $\frac{dT_d}{dT_b} = \frac{T_a T_c}{T_b^2}$ for the derivation into Eq. $\frac{d\,\text{Work}}{dT_b} = 0 = -1 - \frac{dT_d}{dT_b}$ gives the compressor exit temperature for maximum work as $T_b = \sqrt{T_a T_c}$. In terms of temperature ratio, we have

Compressor temperature ratio for maximum work: $\frac{T_b}{T_a} = \sqrt{\frac{T_c}{T_c}}$

The condition for maximum work in a Brayton cycle is different than that for maximum efficiency. The role of the temperature ratio can be seen if we examine the work per unit mass which is delivered at this condition:

$$\text{Work/ Unit Mass} = c_p \left[T_c - \sqrt{T_c T_a} - \frac{T_c T_a}{\sqrt{T_a T_c}} + T_a \right]$$

Rationing all temperatures to the engine inlet temperature:

$$\text{Work/ Unit Mass} = c_p T_a \left[\frac{T_c}{T_a} - 2\sqrt{\frac{T_c}{T_a}} + 1 \right]$$

To find the power the engine can produce, we need to multiply the work per unit mass by the mass flow rate:

$$\text{Power} = \dot{m} c_p T_a \left[\frac{T_c}{T_a} - 2\sqrt{\frac{T_c}{T_a}} + 1 \right]$$

Maximum power for an ideal Brayton cycle:

$$\left(\text{The Units are}: \frac{\text{kg J}}{\text{s kg-K}} \text{K} = \frac{\text{J}}{\text{s}} = \text{Watts} \right)$$

The trend of work output versus compressor pressure ratio, for different temperature ratios, TR = T_c/T_a, is shown in figure.

Figure shows the expression for power of an ideal cycle compared with data from actual jet engines. Figure shows the gas turbine engine layout including the core (compressor, burner, and turbine). Figure shows the core power for a number of different engines as a function of the turbine rotor entry temperature.

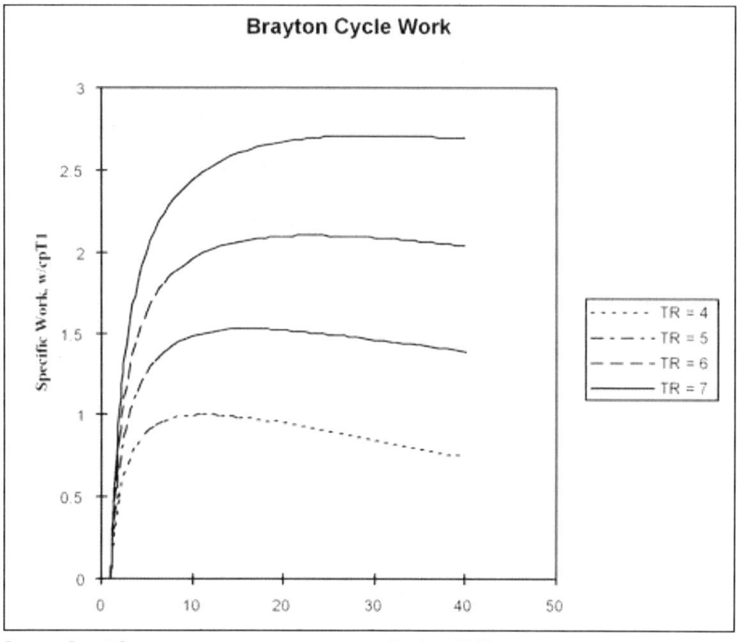

Trend of cycle work with compressor pressure ratio for different temperature ration TR =T_c/T_a.

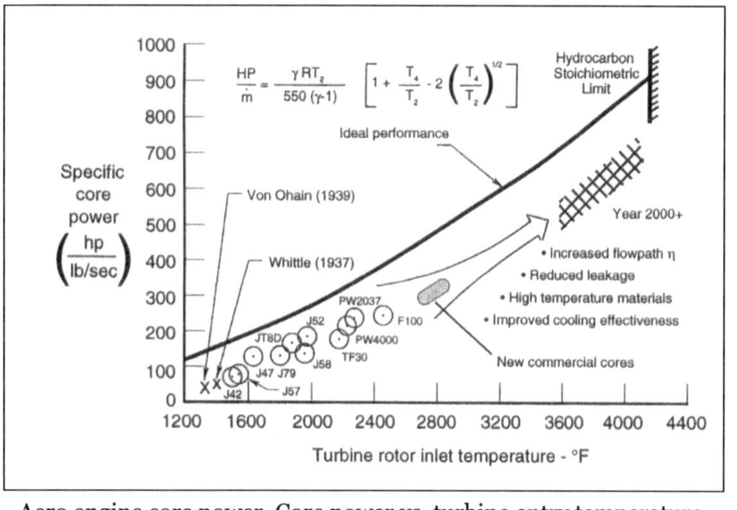

Aero engine core power. Core power vs. turbine entry temperature.

The equation in the figure for horsepower (HP) is the same as that which we just derived, except for the conversion factors. The analysis not only shows the qualitative trend very well but captures much of the quantitative behavior too.

A final comment (for this section) on Brayton cycles concerns the value of the thermal efficiency. The Brayton cycle thermal efficiency contains the ratio of the compressor exit temperature to atmospheric temperature, so that the ratio is not based on the highest temperature in the cycle, as the Carnot efficiency is. For a given maximum cycle temperature, the Brayton cycle is therefore less efficient than a Carnot cycle.

PERFORMANCE AND EFFICIENCY OF GAS TURBINES

The type of operation for which the engine is designed dictates the performance requirement of a gas turbine engine. The performance requirement is mainly determined by the amount of shaft horsepower (s.h.p.) the engine develops for a given set of conditions. The majority of aircraft gas turbine engines are rated at standard day conditions of 59 °F and 29.92 inches Hg. This provides a baseline to which gas turbine engines of all types can be compared.

The need for high efficiency in the engine becomes more important as fuels become more costly. Engine efficiency is primarily defined by the specific fuel consumption (s.f.c.) of the engine at a given set of conditions.

Many factors affect both the efficiency and the performance of the engine. The mass flow rate of air through the engine will dictate engine performance. Any restrictions acting against the smooth flow of air through the engine will limit the engine's performance. The pressure ratio of the compressor, the engine operating temperatures (turbine inlet temperature), and the individual component efficiencies will also influence both the performance and the efficiency of the overall engine. All these factors are considered during the design of the engine. An optimum pressure ratio, turbine inlet temperature, and air mass flow rate are selected to obtain the required performance in the most efficient manner. In addition, individual engine components are designed to minimize flow losses to maximize component efficiencies.

The following graphic shows the typical temperature and pressure rise through the gas flow path.

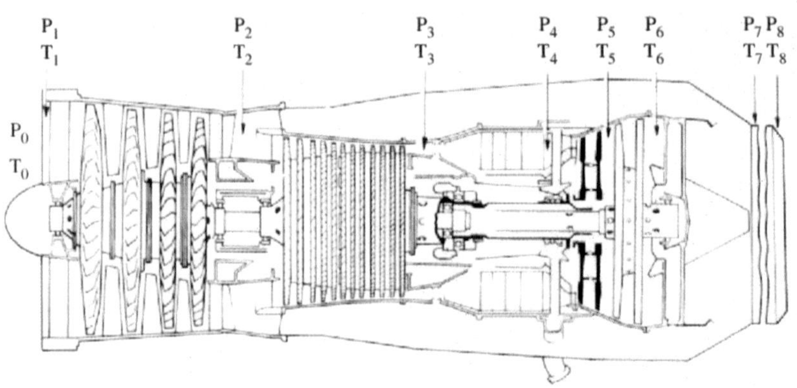

$P_0 T_0$ ambient

$P_1 T_1$ inlet

$P_2 T_2$ low-pressure compressor delivery

$P_3 T_3$ high-pressure compressor delivery

$P_4 T_4$ turbine entry

$P_5 T_5$ high-pressure turbine exit

$P_6 T_6$ low-pressure turbine exit

$P_7 T_7$ exhaust

$P_8 T_8$ propelling nozzle

Engine temperature and pressure flow.

EFFECTS OF GAS TURBINE TEMPERATURE

The materials used in the turbine section of the engine limit the maximum temperature at which a gas turbine engine can operate. The first metal the hot gases from the combustion section strike is the turbine inlet. The temperature of the gas stream is carefully monitored to ensure that over-temperature does not occur.

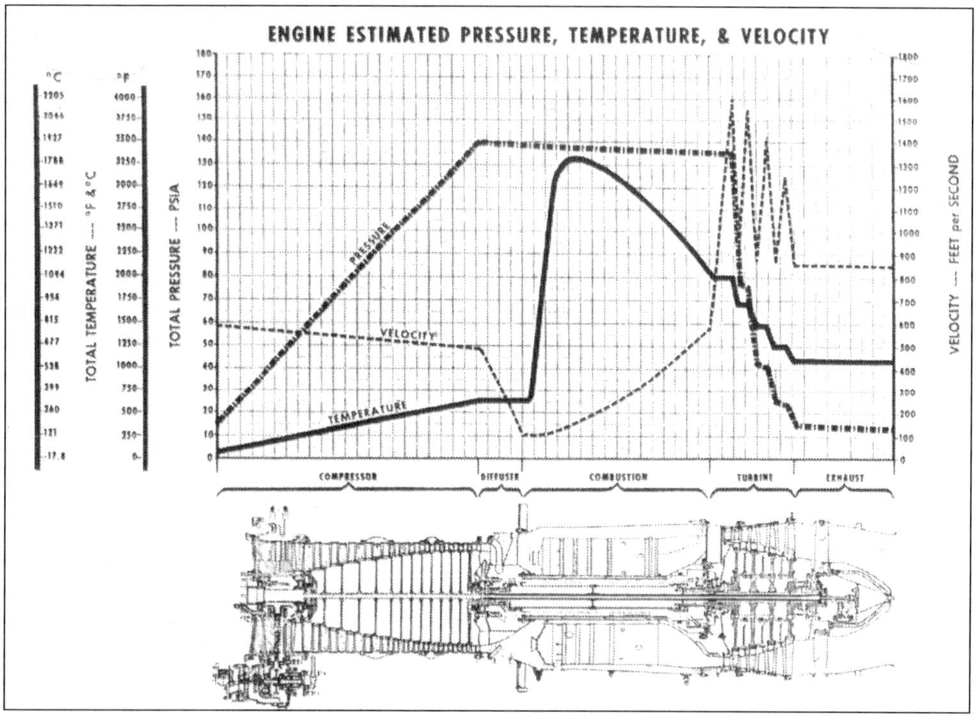

Sample engine pressure, temperature, and velocity.

Compromises are made in turbine design to achieve the optimum balance of power, efficiency, cost, engine life, and other factors. As an example, our sample engine can operate at a higher turbine inlet temperature than previous models due to improved materials and design. The higher temperature allows for increased power and improved efficiency while adding higher cost for the direct cooling of the first turbine stage airfoils and other components.

Advantages and Disadvantages of Gas Turbines

Advantages:

- It requires less space for installation.
- The installation and running cost of gas turbines are less compare to others.
- It has very high power to weight ratio.
- It generates less vibration compare to reciprocating engine.
- It starts easily and quickly.

- It can work in changing load condition easily.
- Its efficiency is higher than IC engines.
- It can develop uniform torque, which is not possible in IC engines.

Disadvantages:

- Starting problem. It cannot start easily because compressor is driven by the turning itself. So an external unit is required to rotate the compressor to start the turbine.
- Most of power is used to drive the compressor so it gives less output.
- Overall efficiency of turbine is low because exhaust gases contain most of heat.

References

- Gas-turbine, encyclopedia: energyeducation.ca, Retrieved 12 May, 2019
- Engine-fundamentals: cast-safety.org, Retrieved 18 August, 2019
- Major-components-of-gas-turbine-engines, technology: britannica.com, Retrieved 25 June, 2019
- Gas-Turbine-Working-Principles- 300857212: researchgate.net, Retrieved 04 January, 2019
- Engine-fundamentals: cast-safety.org, Retrieved 08 April, 2019
- Gas-turbine-types-advantages-disadvantages-working-and-application: mech4study.com, Retrieved 16 July, 2019

Types of Gas Turbines

2 CHAPTER

There are different types of gas turbines which depend on the aircraft engine overhaul. It includes turboprop engines, aeroderivative gas turbines, turboshaft engines, radial gas turbines, microturbines, etc. The topics elaborated in this chapter will help in gaining a better perspective about these types of gas turbines.

JET ENGINES

Jet engines are a class of internal-combustion engines that propel aircraft by means of the rearward discharge of a jet of fluid, usually hot exhaust gases generated by burning fuel with air drawn in from the atmosphere.

Jet engine.

General Characteristics

The prime mover of virtually all jet engines is a gas turbine. Variously called the core, gas producer, gasifier, or gas generator, the gas turbine converts the energy derived from the combustion of a liquid hydrocarbon fuel to mechanical energy in the form of a high-pressure, high-temperature airstream. This energy is then harnessed by what is termed the propulsor (e.g., airplane propeller and helicopter rotor) to generate a thrust with which to propel the aircraft.

Principles of Operation

The Prime Mover

The gas turbine operates on the Brayton cycle in which the working fluid is a continuous flow of air ingested into the engine's inlet. The air is first compressed by a turbocompressor to a pressure ratio of typically 10 to 40 times the pressure of the inlet airstream. It then flows into a combustion chamber, where a steady stream of the hydrocarbon fuel, in the form of liquid spray droplets and vapour or both, is introduced and burned at approximately constant pressure. This gives rise to a continuous stream of high-pressure combustion products whose average temperature is typically from 980 to 1,540 °C or higher. This stream of gases flows through a turbine, which is linked by a torque shaft to the compressor and which extracts energy from the gas stream to drive the compressor. Because heat has been added to the working fluid at high pressure, the gas stream that exits from the gas generator after having been expanded through the turbine contains a considerable amount of surplus energy i.e., gas horsepower by virtue of its high pressure, high temperature, and high velocity, which may be harnessed for propulsion purposes.

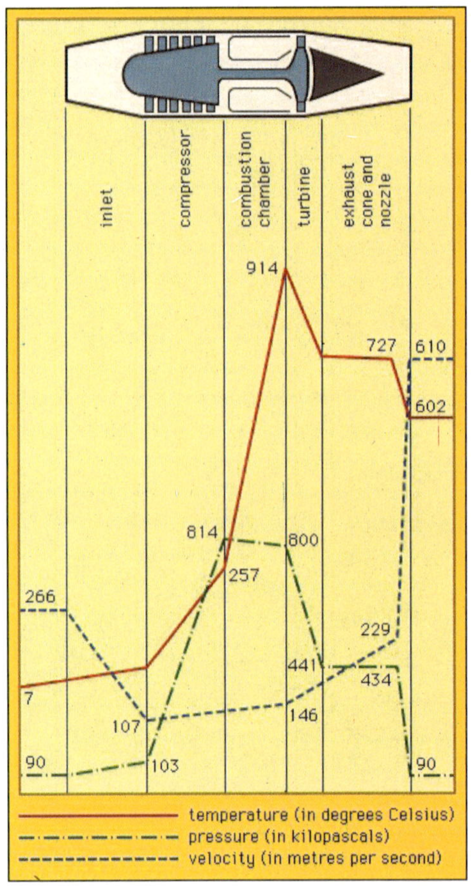

Cross section of a turbojet and (below) graph of typical operating conditions for its working fluid.

The heat released by burning a typical jet fuel in air is approximately 43,370 kilojoules per kilogram (18,650 British thermal units per pound) of fuel. If this process were 100 percent efficient, it would then produce a gas power for every unit of fuel flow of 7.45 horsepower/(pounds per hour), or 12 kilowatts/(kg per hour). In actual fact, certain practical thermodynamic limitations, which are a function of the peak gas temperature achieved in the cycle, restrict the efficiency of the process to about 40 percent of this ideal value. The peak pressure achieved in the cycle also affects the efficiency of energy generation. This implies that the lower limit of specific fuel consumption (SFC) for an engine producing gas horsepower is 0.336 (pound per hour)/horsepower, or 0.207 (kg per hour)/kilowatt. In actual practice, the SFC is even higher than this lower limit because of inefficiencies, losses, and leakages in the individual components of the prime mover.

Because weight and volume are at a premium in the overall design of an aircraft and because the power plant represents a large fraction of any aircraft's total weight and volume, these parameters must be minimized in the engine design. The airflow that passes through an engine is a representative measure of the engine's cross-sectional area and hence its weight and volume. Therefore, an important figure of merit for the prime mover is its specific power the amount of power that it generates per unit of airflow. This quantity is a very strong function of the peak gas temperature in the core at the discharge of the combustion chamber. Modern engines generate from 150 to 250 horsepower/(pound per second), or 247 to 411 kilowatts/(kg per second).

The Propulsor

The gas horsepower generated by the prime mover in the form of hot, high-pressure gas is used to drive the propulsor, enabling it to generate thrust for propelling or lifting the aircraft. The principle on which such a thrust is produced is based on Newton's second law of motion. This law generalizes the observation that the force (F) required to accelerate a discrete mass (m) is proportional to the product of that mass and the acceleration (a). In effect,

$$F = ma = \frac{wa}{g},$$

where the mass is taken as the weight (w) of the object divided by the acceleration due to gravity (g) at the place where the object was weighed. In the case of a jet engine, one is generally dealing with the acceleration of a steady stream of air rather than with a discrete mass. Here, the equivalent statement of the second law of motion is that the force (F) required to increase the velocity of a stream of fluid is proportional to the product of the rate of mass flow (M) of the stream and the change in velocity of the stream,

$$F = M(V_j - V_0) = \frac{W(V_j - V_0)}{g},$$

where the inlet velocity (V_0) relative to the engine is taken to be the flight velocity and the discharge velocity (V_j) is the exhaust or jet velocity relative to the engine. W is the rate of weight flow of working fluid (i.e., air or products of combustion) divided by the acceleration of gravity in the place where the weight flow is measured. The relatively small effect of the weight flow of fuel in creating a difference between the weight flow of the inlet and exhaust streams is intentionally disregarded.

One thereby infers that the components of a propulsor must exert a force F on the stream of air flowing through the propulsor if this device accelerates the airstream from the flight velocity V_0 to the discharge velocity V_j. The reaction to that force F is ultimately transmitted by the mounts of the propulsor to the aircraft as propulsive thrust.

There are two general approaches to converting gas horsepower to propulsive thrust. In one, a second turbine (i.e., a low-pressure, or power, turbine) may be introduced into the engine flow path to extract additional mechanical power from the available gas horsepower. This mechanical power may then be used to drive an external propulsor, such as an airplane propeller or helicopter rotor. In this case, the thrust is developed in the propulsor as it energizes and accelerates the airflow through the propulsor i.e., an airstream separate from that flowing through the prime mover.

In the second approach, the high-energy stream delivered by the prime mover may be fed directly to a jet nozzle, which accelerates the gas stream to a very high velocity as it leaves the engine, as is typified by the turbojet. In this case, the thrust is developed in the components of the prime mover as they energize the gas stream.

In other types of engines, such as the turbofan, thrust is generated by both approaches: A major part of the thrust is derived from the fan, which is powered by a low-pressure turbine and which energizes and accelerates the bypass stream. The remaining part of the total thrust is derived from the core stream, which is exhausted through a jet nozzle.

Just as the prime mover is an imperfect device for converting the heat of fuel combustion to gas horsepower, so the propulsor is an imperfect device for converting the gas horsepower to propulsive thrust. There is generally a great deal of energy left in the high-temperature, high-velocity jet stream exiting from the propulsor that is not fully exploited for propulsion. The efficiency of a propulsor, propulsive efficiency ηp, is the portion of the available energy that is usefully applied in propelling the aircraft compared to the total energy of the jet stream. For the simple but representative case of the discharge airflow equal to the inlet gas flow, it is found that:

$$\eta_p = \frac{2V_0}{V_j + V_0}.$$

Although the jet velocity V_j must be larger than the aircraft velocity V_0 to generate useful thrust, a large jet velocity that exceeds flight speed by a substantial margin can be very detrimental to propulsive efficiency. Maximum propulsive efficiency is approached when the jet velocity is almost equal to (but, of necessity, slightly higher than) the flight speed. This fundamental fact has given rise to a large variety of jet engines, each designed to generate a specific range of jet velocities that matches the range of flight speeds of the aircraft that it is supposed to power.

The net assessment of the efficiency of a jet engine is the measurement of its rate of fuel consumption per unit of thrust generated (e.g., in terms of pounds, or kilograms, per hour of fuel consumed per pounds, or kilograms, of thrust generated). There is no simple generalization of the value of specific fuel consumption of a thrust engine. It is not only a strong function of the prime mover's efficiency (and hence its pressure ratio and peak-cycle temperature) but also of the propulsive efficiency of the propulsor (and hence of the engine type). It also is a strong function of the aircraft flight speed and the ambient temperature (which is in turn a strong function of altitude, season, and latitude).

Basic Engine Types

Achieving a high propulsive efficiency for a jet engine is dependent on designing it so that the exiting jet velocity is not greatly in excess of the flight speed. At the same time, the amount of thrust generated is proportional to that very same velocity excess that must be minimized. This set of restrictive requirements has led to the evolution of a large number of specialized variations of the basic turbojet engine, each tailored to achieve a balance of good fuel efficiency, low weight, and compact size for duty in some band of the flight speed altitude mission spectrum. There are two major general features characteristic of all the different engine types, however. First, in order to achieve a high propulsive efficiency, the jet velocity, or the velocity of the gas stream exiting the propulsor, is matched to the flight speed of the aircraft—slow aircraft have engines with low jet velocities and fast aircraft have engines with high jet velocities. Second, as a result of designing the jet velocity to match the flight speed, the size of the propulsor varies inversely with the flight speed of the aircraft—slow aircraft have very large propulsors, as, for example, the helicopter rotor—and the relative size of the propulsor decreases with increasing design flight speed—turboprop propellers are relatively small and turbofan fans even smaller.

Although the turbojet is the simplest jet engine and was invented and flown first among all the engine types, it seems useful to examine the entire spectrum of engines in the order of the flight-speed band in which they serve, starting with the slowest—namely, the turboshaft engine, which powers helicopters.

Turboshaft Engines

The helicopter is designed to operate for substantial periods of time hovering at zero flight speed. Even in forward flight, helicopters rarely exceed 240 kilometres per hour or a Mach number of 0.22. (The Mach number is the ratio of the velocity of the aircraft to the speed of sound). The principal propulsor is the helicopter rotor, which is driven by one or more turboshaft engines in all modern helicopters of large size. As was previously noted, the propulsor is designed to give a very low discharge or jet velocity and is by the same token very large for a given size aircraft when compared to the propulsors of higher-speed aircraft. The prime mover of a helicopter is a core engine whose gas horsepower is extracted by a power turbine, which then drives the helicopter rotor via a speed-reducing (and combining) gearbox. The power turbine is usually located on a spool separate from the gas generator; thus its rotative speed and that of the helicopter rotor which it drives are independent of the rotative speed of the gas generator. This allows the rotor speed to be varied or kept constant independently of the gas-generator speed, which must be varied to modulate the amount of power generated.

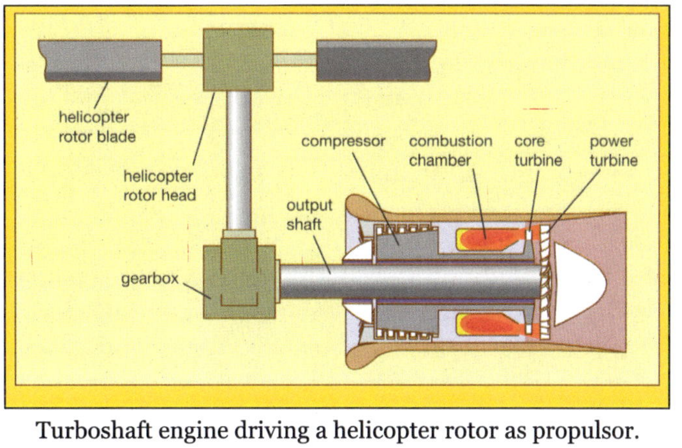

Turboshaft engine driving a helicopter rotor as propulsor.

Turboprops, Propfans and Unducted Fan Engines

The turboprop is the power plant that occupies the next band of flight speeds in the flight spectrum, from a Mach number of 0.2 to 0.7. The propulsor is a propeller with a somewhat higher discharge, or jet velocity, than that of the helicopter rotor to match the flight speed, and it has a proportionately smaller area than the latter for a similarly sized aircraft. The prime mover is a turboshaft engine (very similar to the one that drives a helicopter rotor except for a different gearbox) designed to provide a somewhat higher rotative speed for the propeller, which turns faster than the helicopter rotor having a much larger diameter. The control mode of the turboprop also is somewhat different from that of a helicopter's turboshaft engine. In a helicopter the pilot calls for power by manipulating the pitch of the rotor blades (a greater pitch taking a bigger "bite" of air and so demanding more power to maintain rotative speed). The engine's control responds by increasing fuel to the engine to maintain output shaft speed. In a turboprop the pilot calls for power by selecting fuel flow to the prime mover. The propeller control responds by varying propeller pitch to attain a greater "pull" while maintaining a preselected propeller rotative speed.

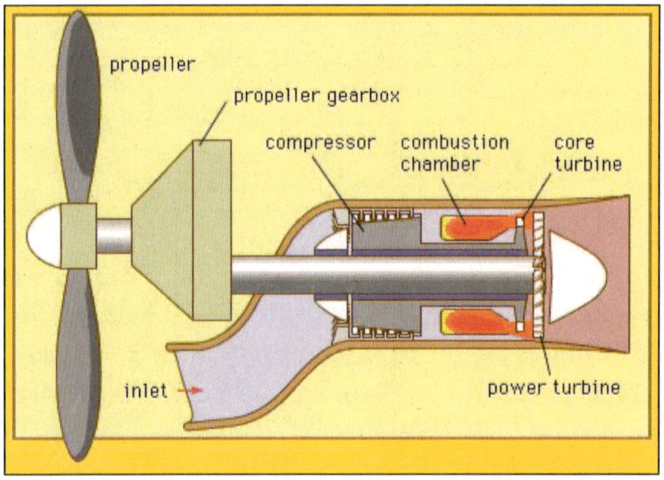

Turboprop engine driving a single rotation propeller as propulsor; tractor arrangement.

A recent trend in turboprop design has been the evolution of propellers for efficient operation at transonic flight speeds (those approximating the speed of sound), much higher than previously achieved—up to Mach numbers of 0.85. This usually involves a higher disk loading (i.e., a higher discharge velocity from the propeller) in order to permit the use of a smaller diameter propeller. This trend has been accompanied by an increase in the number of blades in the propeller (from 6 to 12 instead of the more common 2 to 4 blades in lower-speed propellers). The blades are scimitar-shaped, with swept-back leading edges at the blade tips to accommodate the large Mach numbers encountered by the propeller tip at high rotative and flight speeds. Such high-speed propulsors are called propfans.

Another variation of the propulsor involves the application of two concentric propellers on the same centreline, driven by the same prime mover through a gearbox that causes each propeller to rotate in a direction opposite the other. Such counter-rotating propellers are capable of significantly higher propulsive efficiency and higher disk loading than conventional propellers.

In most turboprop installations the prime mover is mounted on the wing, and the plane of the propeller is forward of the prime mover (the so-called tractor layout). Modern high-speed aircraft may find it more advantageous to mount the engine more toward the rear of the aircraft, with the plane of the propeller aft of the engine. These arrangements are referred to as "pusher" layouts. A recently developed engine layout, identified as the unducted fan (or UDF; trademark), provides a set of very high-efficiency counter-rotating propeller blades, each blade mounted on one of either of two sets of counter-rotating low-pressure turbine stages and achieving all the advantages of the arrangement without the use of a gearbox.

Medium-bypass Turbofans, High-bypass Turbofans and ultrahigh-bypass Engines

Moving up in the spectrum of flight speeds to the transonic regime—Mach numbers from 0.75 to 0.9—the most common engine configurations are turbofan engines, such as those shown in figures Below. In a turbofan, only a part of the gas horsepower generated by the core is extracted to drive a propulsor, which usually consists of a single low-pressure-ratio, shrouded turbocompression stage. The fan is generally placed in front of the core inlet so that the air entering the core first passes through the fan and is partially compressed by it. Most of the air, however, bypasses the core (hence the designation bypass stream) and goes directly to an exhaust nozzle. The core stream, with some modest fraction of the gas horsepower remaining (not extracted to drive the fan) proceeds directly to its own exhaust nozzle.

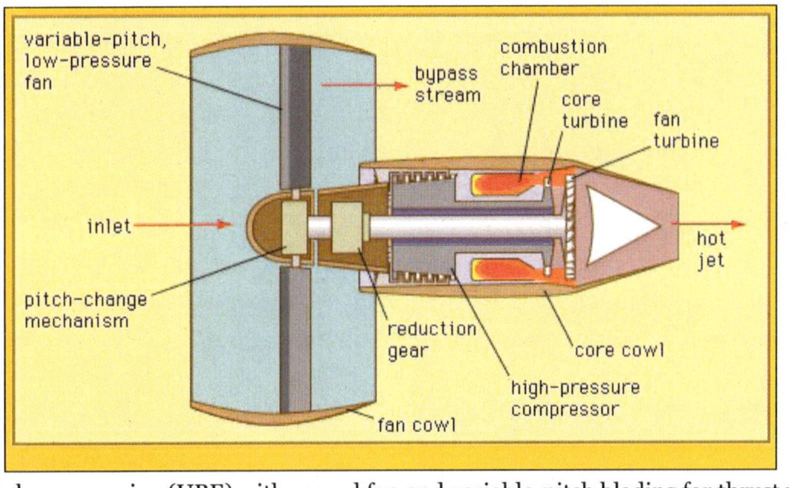

Ultrahigh-bypass engine (UBE) with geared fan and variable-pitch blading for thrust reversal.

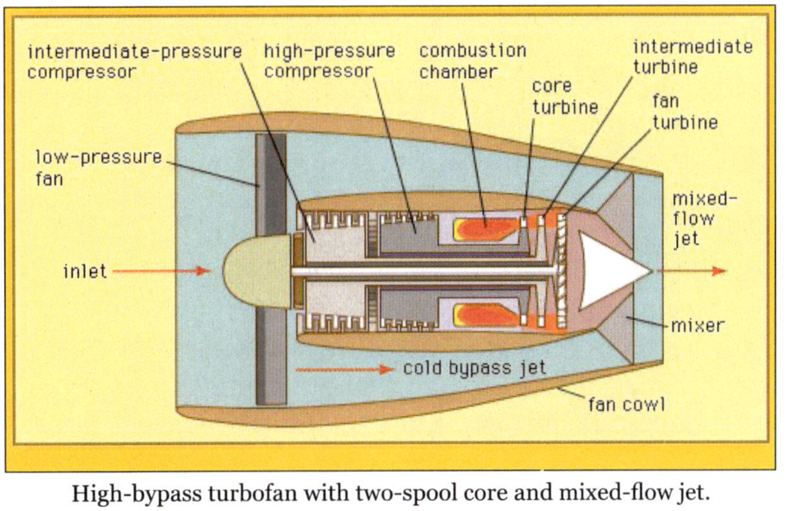

High-bypass turbofan with two-spool core and mixed-flow jet.

A key parameter for classifying the turbofan is its bypass ratio, defined as the ratio of the mass flow rate of the bypass stream to the mass flow rate entering the core. Since the highest propulsion efficiencies are obtained by the engines with the highest bypass ratios, one would expect to find all engines of that design in this flight speed regime. (Some of the variation derives from historical evolution). In actuality, however, one finds engines with a broad spectrum of bypass ratios, including medium-bypass engines (with bypass ratios from 2 to 4), high-bypass engines (with bypass ratios from 5 to 8), and ultrahigh-bypass engines, so-called UBEs (with bypass ratios from 9 to 15 or higher). A whole generation of low- and medium-bypass engines has completely supplanted the first generation of aircraft powered by (zero-bypass) turbojet engines. Moreover, that generation was itself supplanted by a third generation of medium- and high-bypass turbofan engines. There are several other reasons why engines with less than the highest bypass ratios hypothetically achievable are still in use. Very high bypass ratios involve the use of fans with very large diameters, which in turn entail very heavy components; this increases the difficulty of installing the engine on aircraft and maintaining sufficient ground clearance. In addition, the weight and complexity of the apparatus required to reverse the direction of the bypass stream (to achieve thrust reversal in order to shorten the aircraft's landing roll) also increases with the bypass ratio. The long-term trend, however, is definitely toward higher and higher bypass ratios.

There are several unique features and ancillary devices found in turbofan engines. Ultrahigh-bypass engines may have a gearbox between the drive turbine and the fan to simplify the design of the small-diameter turbine (with the attendant high rotative speed) without compromising the performance of the very large-diameter fan (with the attendant low rotative speed). Variable-pitch fan blades are generally required for thrust reversal in such ultrahigh-bypass fans, while in medium- and high-bypass engines the thrust reversing is usually accomplished by introducing blocker doors into the bypass stream. In high- and medium-bypass turbofans, a small but significant improvement in propulsive efficiency can be achieved by mixing the airstream of the hot core and cold bypass streams before the total airstream enters a single jet nozzle.

Low-bypass Turbofans and Turbojets

In the next higher regime of aircraft flight speed, the low supersonic range from Mach numbers above 1 up to 2 or 3, one finds the application of the simple turbojet (with no bypass stream) and the low-bypass turbofan engine (with a bypass ratio up to 2).

Although the low-bypass turbofan has the same general appearance as a turbofan with a larger bypass ratio, certain special features are unique to low-bypass engines. The lower total flow in the fan generally involves a higher fan pressure ratio (for equivalent amounts of energy available from the drive turbine), and so such a fan usually has more than one (i.e., two or three) turbocompressor stages. Engines designed to operate at the low supersonic range generally have insufficient thrust in other flight regimes or modes where they must operate for short durations, as, for instance, acceleration through transonic speed, takeoff from high-altitude airports under conditions of extremely high temperatures and high gross weight, or combat maneuvers at high supersonic flight speed. Rather than installing a larger engine to meet these requirements, it is more effective to add an afterburner to a turbofan engine as a means of thrust augmentation. The afterburner is a secondary combustion system that operates in the exhaust stream of the engine before the stream is introduced into the exhaust nozzle. Such a device is not as fuel-efficient as the main turbofan section of the engine because heat addition occurs at a lower pressure than in the main burner. The afterburner, however, is relatively simple and lightweight, since it does not contain any rotating machinery. For the same reason, it may be operated to a much higher discharge temperature (typically 1,760 °C), so that it is capable of augmenting the thrust of the turbofan by as much as 50 percent.

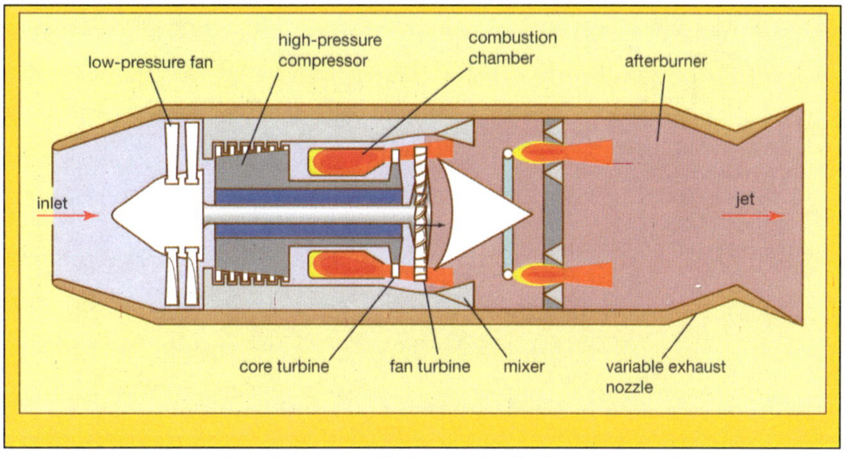

Low-bypass turbofan with afterburner.

The afterburner in a turbofan usually requires a mixer for mixing the relatively cool bypass air with the hot core stream; the cooler air is otherwise difficult to burn in the low-pressure environment of an afterburner. Also, in both the turbojet and the turbofan with an afterburner, the exhaust nozzle must have a variable throat area to accommodate the large variations in volumetric flow rate between the very hot exhaust stream from the operating afterburner and the cooler airstream discharged from the engine when the afterburner is not in use. Engines intended for supersonic flight generally have a much lower compression-pressure ratio than higher-bypass machines intended for subsonic or transonic operation. A major contributor to this tendency is the additional pressure ratio developed in the engine's inlet as it slows down or diffuses the very high-speed airstream that is ingested as the engine's working fluid—the ram effect. At transonic flight speed this pressure ratio is almost 2:1, so that the engine's compressor may be built to provide that much less pressure where peak pressure is otherwise limiting.

Early generations of jet-propelled aircraft in this low supersonic flight regime were powered by turbojet engines, but subsequent generations built for the same flight regime have largely been equipped with low-bypass turbofans. This substitution of engine type was undertaken primarily because such aircraft expend a great deal of their fuel at subsonic flight speed (e.g., in takeoff, climb, loiter, acceleration, approach, and landing), where the turbofan provides an advantage in propulsive efficiency.

Ramjets and Supersonic Combustion Ramjets

As has been seen, ram pressure plays an increasingly important role in the thermodynamic cycle of power and thrust generation of the jet engine at supersonic flight speeds. For flight speeds above Mach 2.5 or 3, the ram-pressure ratio becomes so high that a turbocompressor is no longer necessary for efficient thrust generation. Indeed, the pressure ratio eventually rises to such high values that the associated high ram temperatures make it difficult or impossible to place high-speed rotating machinery in the flow path without prohibitive amounts of cooling provision. This combination of circumstances gives rise to the ramjet, a jet engine in which the pressure increase is attributable only to the ram effect of the high flight speed; no turbomachinery is involved, and the main thrust producer is an afterburner.

Ramjets are lightweight and simple power plants, making them ideal candidates for supersonic flight vehicles that are launched from other flight vehicles at extremely high speed. They are less suitable for use in vehicles that must be sufficiently self-powered for subsonic takeoff, climb, and acceleration to supersonic flight speed; the subsonic ram pressure is insufficient to produce any reasonable amount of thrust, and so alternative propulsion devices must be provided.

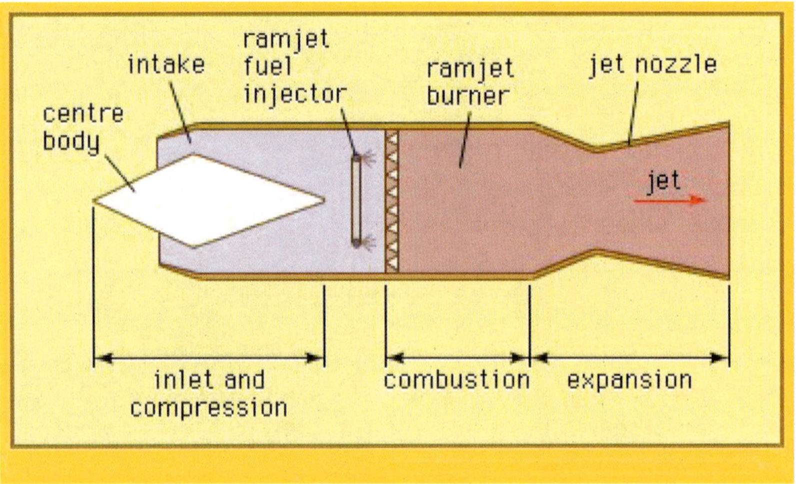

Arrangement of a ramjet.

In the flight regime of Mach 4 or 5, it is usually efficient to decelerate the inlet airstream to subsonic velocity before it enters the combustion system. At still higher Mach numbers, such deceleration becomes more difficult and costly in terms of pressure losses, and it is necessary to make provision for the combustion chamber to burn its fuel in the supersonic airstream. Such specialized ramjets are called scramjets (for supersonic combustion ramjets) and are projected to be fueled by a cryogenically liquified gas (e.g., hydrogen or methane) instead of a liquid hydrocarbon. The primary reason for doing so is to exploit the greater heat release per unit weight of fuels that have a higher ratio of hydrogen to carbon atoms than ordinary fractions of petroleum even though this gain is partly negated by the higher volume per unit of heat release of those same fuels. Another incentive for employing a very cold fuel is that it may be used as a heat sink for cooling a very high-speed (and hence very hot) engine and aircraft structure. The scramjet has an unusual feature: the inlet deceleration and exhaust acceleration occur largely outside the enclosed engine inlet and exhaust ducts against external aircraft surfaces in front of and to the rear of the engine. In effect, the engine itself is little more than a sophisticated supersonic combustion chamber.

Hybrid Engine Types

It is possible to tailor an engine configuration so that the engine is well suited for operation within a given band of the flight spectrum. To have an engine that will perform well in more than one band of the flight spectrum or in more than one regime of operation, it may be necessary to configure the power plant so that it can be converted from one engine type to another by means of variable geometry built into the engine components.

Vertical and Short Takeoff and Landing (V/STOL) Propulsion Systems

Propulsion systems that provide aircraft with the capability of both vertical and conventional forward flight represent a formidable challenge to the engine designer. V/STOL aircraft have several major categories of engine arrangement. They are as follows:

- As in a helicopter, the propulsor may consist of a rotor that is driven by one or more turboshaft engines and is installed in such a way as to provide vertical thrust. The entire aircraft must be tilted to give the thrust vector a forward component to achieve forward flight. This arrangement has certain limitations in terms of effectiveness, as borne out by the relative inefficiency of forward flight above a Mach number of 0.2.

- The propulsors may be mounted on pivots so that they can be rotated from the position in which they give vertical thrust in a takeoff, hover, climb, descent, or landing maneuver and pivoted 90° to provide thrust for conventional forward flight (as in the tilt-rotor aircraft). The prime mover that drives the propulsor may either be tilted with the propulsor or be fixed in the wing and drive the tilting propulsor via a rotating shaft through the pivot axis. In some configurations, the entire wing of the aircraft, carrying fixed engines and propulsors, may be tilted as a single assembly.

- The engines may be fixed in a position required to produce thrust for forward flight. Their exhaust systems, however, have built-in variable geometry, making it possible to vector the exhaust nozzle (or nozzles) or divert the exhaust gases by means of valves and auxiliary ducts to nozzles mounted in such a way as to provide vertical thrust or lift.

- The aircraft may include two different sets of engines or propulsors (or both), fixed in position, with one set installed for forward flight and the other for vertical thrust (i.e., the lift engines).

- The aircraft may use a convertible engine. Such an engine has a single prime mover that is arranged to drive a fan for efficient forward propulsion, to drive a shaft that turns the main helicopter rotor, or to drive both a fan and a shaft. In order to convert from horizontal to vertical flight, variable-pitch fan blades or variable-pitch stators (or both) unload the fan, thereby making mechanical power available to drive the helicopter rotor for vertical movement.

Variable-cycle Engines

For aircraft designed to fly mixed missions (i.e., at subsonic, transonic, and supersonic flight speeds) with low levels of fuel consumption, it is desirable to have an engine with the characteristics of both a high-bypass engine (for subsonic flight speed) and a low-bypass engine (for supersonic flight speed). This requirement is typical for many high-speed commercial airliners, including the Concorde, a type of supersonic transport built by the British and French that was in service from 1976 to 2003. The Concorde was capable of traveling over oceans and unpopulated land areas at supersonic cruise speeds, but it could not fly efficiently and quietly at subsonic flight speed for takeoff, ascent, cruising over populated areas, and approach and landing. This dual function is expected to be accomplished in the future by the variable-cycle engine (VCE). If the components of an engine are designed to accommodate the extreme limits of flow, pressure ratio, and other conditions involved in both high-bypass and low-bypass operation, the engine may be operated at either extreme of bypass ratio or at any bypass ratio between those extremes by means of a valve (or valves) in the bypass stream (in conjunction with a variable exhaust nozzle). When the valves are closed, they restrict the flow in the bypass stream to achieve low bypass for supersonic flight. When the valves are open, the bypass is increased to its maximum value for efficient subsonic flight.

Turboramjets

The ramjet provides a simple and efficient means of propulsion for aircraft at relatively high supersonic flight speeds. It is, however, quite inefficient at transonic flight speeds and is completely ineffective at subsonic velocities. The turboramjet has been developed to overcome this inadequacy. In this system, a turbofan engine is built into the inlet of a ramjet engine to charge the latter with a pressurized stream of air at subsonic flight speed where ram pressure is insufficient for effective ramjet operation. During supersonic flight the fan blades, if they are of variable pitch, may be feathered so that they do not interfere with the flow of ram air to the ramjet. A separate inlet to the core engine that drives the fan may be closed off so as not to expose the turbomachinery to the hostile environment of the high-temperature ram air.

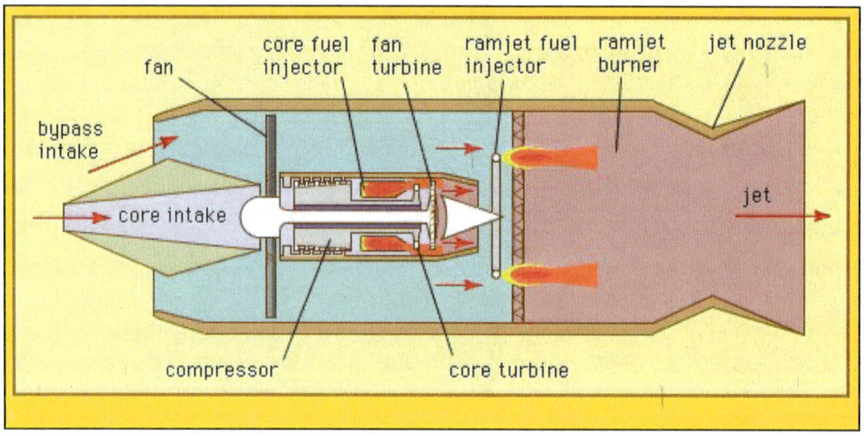

Turboramjet with air-breathing prime mover.

Another variation of the turboramjet does without the core inlet and the core compressor altogether. Instead, the aircraft carries a tank of an oxidizer, such as liquid oxygen. The oxidizer is fed into the core combustion chamber along with the fuel to support the combustion process, which generates the hot gas stream to power the turbine that drives the fan. During supersonic flight, the fan may be feathered, and a surplus of fuel may be introduced into the core combustor. The unburned fuel passes through the fan turbine and undergoes combustion in the ramjet burner when it mixes with the fresh air entering via the bypass stream from the fan.

Auxiliary Power Units

APUs are small gas turbines designed to supply auxiliary power to larger, mobile, machines such as an aircraft. They supply:

- Compressed air for air conditioning and ventilation,
- Compressed air start-up power for larger jet engines,
- Mechanical (shaft) power to a gearbox to drive shafted accessories or to start large jet engines, and

- Electrical, hydraulic and other power-transmission sources to consuming devices remote from the APU.

Working Principle of Auxiliary Power Unit

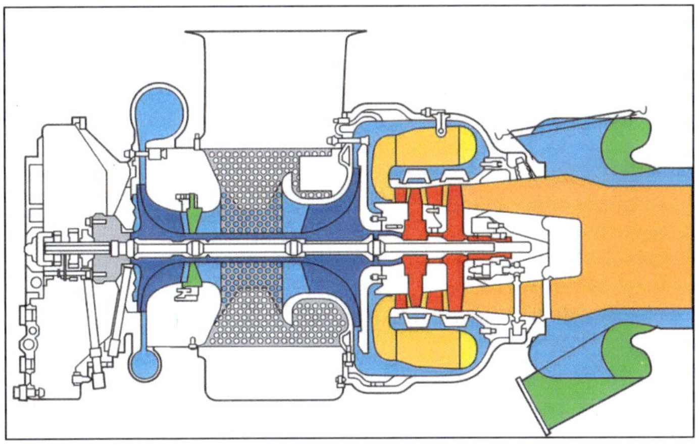

The APU is a small turbine engine installed near the rear of the fuselage.

An aircraft auxiliary power unit serves as an additional energy source normally used to start one of the main engines on an airliner or business jet. The APU is equipped with an extra electrical generator to create enough power to operate onboard lighting, galley electrics and cockpit avionics, usually while the aircraft is parked at the gate. Drawing bleed air from its own compressor, an APU also drives the environmental packs used to heat and cool the aircraft.

And most important, operating an APU negates the need to start one of the aircraft's main engines while waiting for passengers to arrive, thereby saving on fuel and maintenance for a more expensive power plant.

In most cases, the APU is shut down before takeoff and reignited when the aircraft clears the runway after landing. While most of an APU's active service life occurs as the aircraft sits on the ground, in some instances the APU is used as an emergency electrical power source while the aircraft is airborne.

The APU is a small turbine engine installed near the rear of the fuselage. But calling the APU an extra jet engine is not accurate because the turbine exhaust from the APU is vented overboard. A jet engine would be used to propel the aircraft forward.

The earliest APUs could be found on the B-29 Superfortress, looking essentially like a motorcycle engine installed inside the fuselage. The Convair XP5Y-1 also used an early APU, while America's first jet airliner, the Boeing 707, was delivered without one. The 727 was the first Boeing to be APU-equipped.

Today, APUs can be found in medium-size and larger civil and military jets, some turboprop aircraft and a handful of military fighters. Smaller civilian jets like the Cessna Citation CJ or One Aviation's Eclipse jet don't carry an APU because the extra weight of even a small extra turbine engine can significantly impact the airplane's useful load.

The aircraft manufacturer determines APU requirements after considering the size of the cabin, the amount of bleed air required to power the environmental packs, and the generator size needed to power the cockpit and cabin and start an engine. It is then the job of the APU manufacturer, usually a subcontractor to the aircraft manufacturer, to deliver a unit meeting those specifications.

While readying the avionics is an important element of preparing an aircraft for departure, creating a comfortable cabin environment before passengers arrive is almost more important to most operators. During hot summer months, the APU driving the environmental packs might require half an hour to cool the cabin of an aircraft sitting in Miami, and equally as long to heat one in winter on the ramp in Fargo, North Dakota.

During some extended-range twin-engine (ETOPS) flights, which permit twin-engine aircraft to operate on routes more than 60 flying minutes away from a suitable emergency airport, the APU system must be tested — using a cold-start procedure — to verify start reliability in case it might be needed, should one of the aircraft's main engines fail in flight.

One drawback to APUs is the noise they often produce while running on the ground. In some older aircraft, the noise of the APU's turbine engine can become worse once the bleed air is switched on to heat or cool the cabin. A number of airports in the United States still restrict the operation of APUs, especially at night, to avoid annoying people in nearby communities.

Like all modern turbine engines, an APU is equipped with a fire-extinguishing system, often operated automatically in newer aircraft. Those operating an APU without automatic extinguishing systems usually require at least one pilot to remain near the aircraft in order to hop aboard and manually discharge the APU fire bottle should a blaze erupt.

Unlike the aircraft's main engines, which require regular tear-downs at very specific time points for maintenance, many APUs are treated more like pass/fail items, allowing operators to run them until something breaks, unless, of course, the unit is part of an ETOPS operation.

In days past, when the price of jet fuel was considerably more expensive than today, some airlines were known to wait until just before departure time to start the APU in order to save money. Luckily, for passengers at least, that strategy to economize was quickly superseded by passengers' demand to board an aircraft with a comfortable cabin.

TURBOPROP ENGINES

A turboprop engine is a turbine engine that drives an aircraft propeller.

In its simplest form a turboprop consists of an intake, compressor, combustor, turbine, and a propelling nozzle. Air is drawn into the intake and compressed by the compressor. Fuel is then added to the compressed air in the combustor, where the fuel-air mixture then combusts. The hot combustion gases expand through the turbine. Some of the power generated by the turbine is used to drive the compressor. The rest is transmitted through the reduction gearing to the propeller. Further expansion of the gases occurs in the propelling nozzle, where the gases exhaust to atmospheric pressure. The propelling nozzle provides a relatively small proportion of the thrust generated by a turboprop.

GE T64 propeller at left, gearbox with accessories in the middle, turbine core at right.

In contrast to a turbojet, the engine's exhaust gases do not generally contain enough energy to create significant thrust, since almost all of the engine's power is used to drive the propeller.

Technological Aspects

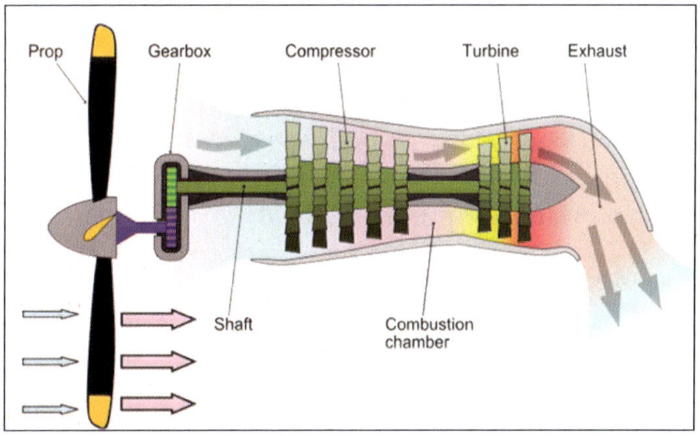

Schematic diagram showing the operation of a turboprop engine.

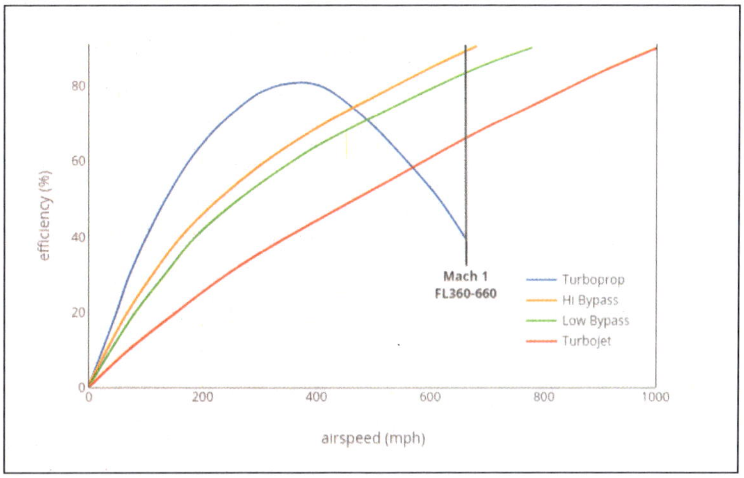

Propulsive efficiency comparison for various gas turbine engine configurations.

Exhaust thrust in a turboprop is sacrificed in favour of shaft power, which is obtained by extracting additional power (up to that necessary to drive the compressor) from turbine expansion. Owing to the additional expansion in the turbine system, the residual energy in the exhaust jet is low. Consequently, the exhaust jet typically produces around or less than 10% of the total thrust. A higher proportion of the thrust comes from the propeller at low speeds and less at higher speeds.

Turboprops can have bypass ratios up to 50-100 although the propulsion airflow is less clearly defined for propellers than for fans.

The propeller is coupled to the turbine through a reduction gear that converts the high RPM/low torque output to low RPM/high torque. The propeller itself is normally a constant-speed (variable pitch) propeller type similar to that used with larger aircraft reciprocating engines.

Unlike the small diameter fans used in turbofan jet engines, the propeller has a large diameter that lets it accelerate a large volume of air. This permits a lower airstream velocity for a given amount of thrust. As it is more efficient at low speeds to accelerate a large amount of air by a small degree than a small amount of air by a large degree,a low disc loading (thrust per disc area) increases the aircraft's energy efficiency, and this reduces the fuel use.

Propellers lose efficiency as aircraft speed increases, so turboprops are normally not used on high-speed aircraft above 0.6-0.7 Mach· However, propfan engines, which are very similar to turboprop engines, can cruise at flight speeds approaching 0.75 Mach. To increase propeller efficiency, a mechanism can be used to alter their pitch relative to the airspeed. A variable-pitch propeller, also called a controllable-pitch propeller, can also be used to generate negative thrust while decelerating on the runway. Additionally, in the event of an engine failure, the pitch can be adjusted to a variable pitch (called feathering), thus minimizing the drag of the non-functioning propeller.

While most modern turbojet and turbofan engines use axial-flow compressors, turboprop engines usually contain at least one stage of centrifugal compressor which have the advantage of being simple and lightweight, at the expense of a streamlined shape.

While the power turbine may be integral with the gas generator section, many turboprops today feature a free power turbine on a separate coaxial shaft. This enables the propeller to rotate freely, independent of compressor speed. Residual thrust on a turboshaft is avoided by further expansion in the turbine system and/or truncating and turning the exhaust 180 degrees, to produce two opposing jets. Apart from the above, there is very little difference between a turboprop and a turboshaft.

Usage

A military transport aircraft, over 2,500 Lockheed C-130 Hercules have been built.

Compared to turbofans, turboprops are most efficient at flight speeds below 725 km/h (450 mph; 390 knots) because the jet velocity of the propeller (and exhaust) is relatively low. Modern turboprop airliners operate at nearly the same speed as small regional jet airliners but burn two-thirds of the fuel per passenger. However, compared to a turbojet (which can fly at high altitude for enhanced speed and fuel efficiency) a propeller aircraft has a lower ceiling.

The Beech King Air and Super King Air are the most-delivered turboprop business aircraft in history.

Compared to piston engines, their greater power-to-weight ratio (implying shorter takeoffs) and reliability can offset their higher initial cost, maintenance and fuel consumption. As jet fuel can be easier to obtain than avgas in remote areas, turboprop-powered aircraft like the Cessna Caravan and Quest Kodiak are used as bush airplanes.

Turboprop engines are generally used on small subsonic aircraft, but the Tupolev Tu-114 can reach 470 kt (870 km/h, 541 mph). Large military and civil aircraft, such as the Lockheed L-188 Electra and the Tupolev Tu-95, have also used turboprop power. The Airbus A400M is powered by four Europrop TP400 engines, which are the second most powerful turboprop engines ever produced, after the eleven megawatt-output Kuznetsov NK-12.

In 2017, The most widespread turboprop airliners in service were the ATR 42/72 (950 aircraft), Bombardier Q400 (506) and Dash 8-100/200/300 (374), Beechcraft 1900 (328), de Havilland Canada DHC-6 Twin Otter (270), Saab 340 (225). Less widespread and older airliners include the BAe Jetstream 31, Embraer EMB 120 Brasilia, Fairchild Swearingen Metroliner, Dornier 328, Saab 2000, Xian MA60, MA600 and MA700, Fokker 27 and 50.

Turboprop business aircraft include the Piper Meridian, Socata TBM, Pilatus PC-12, Piaggio P.180 Avanti, Beechcraft King Air and Super King Air. In April 2017, there were 14,311 business turboprops in the worldwide fleet.

Reliability

Between 2012 and 2016, the ATSB observed 417 events with turboprop aircraft, 83 per year, over 1.4 million flight hours: 2.2 per 10,000 hours. Three were "high risk" involving engine malfunction and unplanned landing in singleengine Cessna 208 Caravans, four "medium risk" and 96% "low risk". Two occurrences resulted in minor injuries due to engine malfunction and terrain collision in agricultural aircraft and five accidents involved aerial work: four in agriculture and one in an air ambulance.

AERODERIVATIVE GAS TURBINES

For more than a half-century gas turbine engines pioneered for military jet fighters have hung under the wings of commercial airliners. For nearly as long, manufacturers have built industrial gas turbines to drive electricity generators and pump oil and gas. Many gas turbines, large and small, are designed specifically for these industrial applications. Aeroderivative gas turbines used for these industrial applications are adapted directly from existing aircraft engines. Aeroderivative gas turbines emerged in the late 1960s with unique performance attributes in comparison to the existing industrial gas turbines. Aeroderivative units could startup more quickly for peak and emergency electricity generation. Also, aeroderivative turbines offered lower weight in a smaller footprint, which was ideal for offshore platforms. Furthermore, their higher efficiency, coupled with simplified installation and maintenance, saved money for pipeline operators.

In the mid-1980s independent power producers began using aeroderivative gas turbines for combined heat and power generation, also called cogeneration. In this configuration, the exhaust heat of the gas turbine is used to produce steam to directly heat a building or industrial process. Aeroderivative gas turbines can convert 40 percent of fuel energy into electricity; when configured for cogeneration, system efficiency can exceed 80 percent, as far less of the fuel's chemical energy is lost as unused heat. Aeroderivative gas turbines are also being used to balance the integration of variable power sources, like wind and solar, into the electricity grid.

Most electricity generated by natural gas come from large industrial gas turbines designed specifically for that purpose, not aeroderivative units. These larger, heavy-duty or frame-type gas turbines for baseload power can exceed 200 megawatts (MW) each, and large plants may configure several of these units together in combined cycle. In this configuration, exhaust heat from one or more gas turbines drives a separate steam turbine to generate additional electricity, achieving an overall efficiency around 60 percent.

Aeroderivative gas turbines have historically been limited to midsize units of roughly 18 to 65 MW each, reflecting the size of their parent aircraft engines. While the capacity of aeroderivative units is smaller, these units add flexibility critical for sustaining large generating plants throughout the electricity grid. Moreover, aeroderivative units are well suited for highly efficient cogeneration, more flexible combined cycle plants, and in mechanical drive applications essential to production and distribution of oil and gas.

Worldwide, aeroderivative units account for approximately 10 to 20 percent of gas turbine capacity in recent years' orders, totaling roughly 6,000 to 8,000 MW per year. However, for mid-size units with capacities of 18 to 65 MW, aeroderivative units account for approximately two thirds of gas turbine capacity sold in recent years. General Electric (GE), the leading provider of aeroderivative gas turbines, estimates having over 2,300 aeroderivative units in service for electricity generation worldwide, totaling 80,000 MW of capacity. Rolls Royce projects that aeroderivative units used for electricity generation and oil and gas operations over the next 20 years will be worth $70 billion in sales plus another $50 billion for associated services.

The entire industrial gas turbine market is just over half the size of the commercial and defense aircraft gas turbine engine market, which in 2010 amounted to more than $26 billion worldwide.

The performance of aeroderivative gas turbines is largely enabled by the engineering successes sustained in the lucrative aircraft engine market. The role and impact of industry-government partnerships in advancing aircraft engine technology is widely appreciated and has directly impacted the evolution of aeroderivative gas turbines.

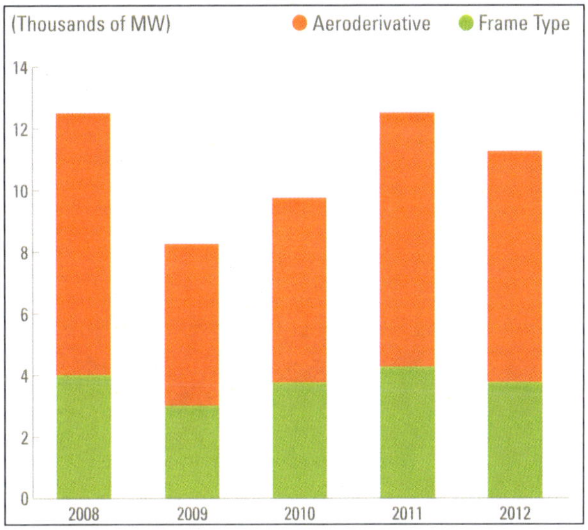

Worldwide gas turbine orders by type, midsize (18-65 MW) only.

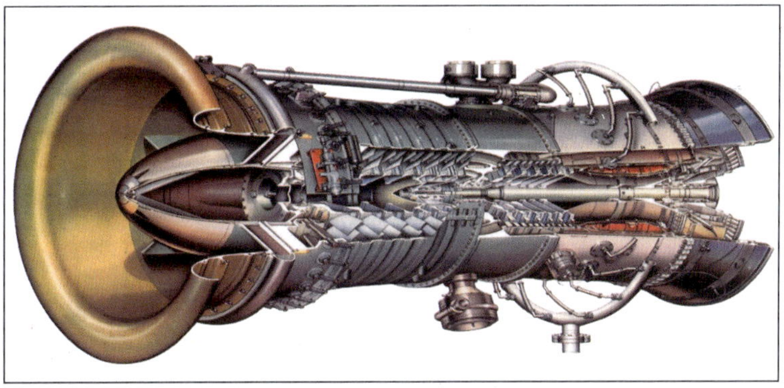

Aeroderivative Gas Turbine Cutaway.

Compressor

Air moves through several stages of rotating blades and is pressurized and thereby heated. Higher pressure ratios between the air inlet and the combustor increases power and simple cycle efficiency.

Combustor

Fuel is burned in the combustor, heating the gas flow to the material temperature limit of the turbine blades. Materials and designs capable of higher temperatures increases power and efficiency.

Turbine

The expanding hot gas rotates the turbine blades and the shaft connecting the turbine to the compressor, continuing the cycle. Heat remaining in the exhaust can be captured for cogeneration or combined cycle.

Much of the core aircraft engine research and development (R&D) has doubled effectively as aeroderivative gas turbine R&D, as the most difficult and costly to develop components in aircraft engines are used in nearly identical form in aeroderivative gas turbines. For example, over the last decade, spending on aircraft engine R&D at GE has exceeded $1 billion per year. By comparison, over the last four decades, spending on aeroderivative gas turbine R&D at GE is estimated somewhere above $2 billion in total. GE, Rolls Royce, and Pratt & Whitney are the three manufacturers of aeroderivative gas turbines, and each of their aeroderivative models is descendent from a specific aircraft engine in their lineup. As a result, tremendous gains in aircraft engine performance, led by the military and bolstered by the rise of commercial aviation, have resulted directly in improvements for aeroderivative gas turbines.

Working Principle of Aeroderivative Gas Turbines

Aeroderivative gas turbines, like all gas turbines, use a continuous intake of air and a continuous injection of fuel to create a hot, pressurized gas flow that expands through the turbine. In the compressor, a series of rotating blades pressurizes the incoming air in stages; this pressurization heats the air. In the combustor, chemical energy from the burning fuel adds far more heat. The hot, pressurized gas expands through the turbine blades and rotates the shaft that drives the compressor at the front of the engine, continuing the cycle. From this basic configuration, the remaining energy not used to drive the compressor can be captured in useful ways for various applications. In aircraft engines, the hot exhaust gas passes through additional turbines to rotate a shaft driving a propeller or fan that provides most of the aircraft's thrust. Aeroderivative and other industrial gas turbines work in a similar way, adding more turbines to extract energy from the hot exhaust gas to power a shaft. For industrial applications, the shaft is connected to an external electricity generator, a pump, or a ship's drivetrain.

TURBOSHAFT ENGINES

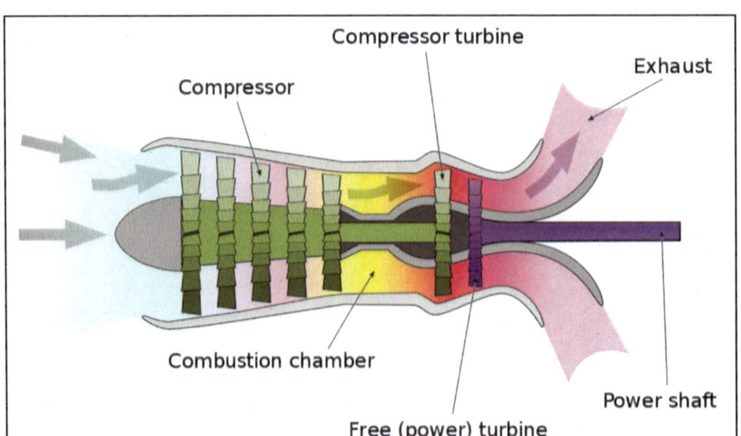

Schematic diagram showing the operation of a simplified turboshaft engine.
The compressor spool is shown in green and the free / power spool is in purple.

A turboshaft engine is a form of gas turbine that is optimized to produce shaft power rather than jet thrust.

In concept, turboshaft engines are very similar to turbojets, with additional turbine expansion to extract heat energy from the exhaust and convert it into output shaft power. They are even more similar to turboprops, with only minor differences, and a single engine is often sold in both forms.

Turboshaft engines are commonly used in applications that require a sustained high power output, high reliability, small size, and light weight. These include helicopters, auxiliary power units, boats and ships, tanks, hovercraft, and stationary equipment.

A turboshaft engine may be made up of two major parts assemblies: the 'gas generator' and the 'power section'. The gas generator consists of the compressor, combustion chambers with ignitors and fuel nozzles, and one or more stages of turbine. The power section consists of additional stages of turbines, a gear reduction system, and the shaft output. The gas generator creates the hot expanding gases to drive the power section. Depending on the design, the engine accessories may be driven either by the gas generator or by the power section.

In most designs, the gas generator and power section are mechanically separate so they can each rotate at different speeds appropriate for the conditions, referred to as a 'free power turbine'. A free power turbine can be an extremely useful design feature for vehicles, as it allows the design to forgo the weight and cost of complex multiple-ratio transmissions and clutches.

The general layout of a turboshaft is similar to that of a turboprop. The main difference is a turboprop is structurally designed to support the loads created by a rotating propeller, as the propeller is not attached to anything but the engine itself. In contrast, turboshaft engines usually drive a transmission which is not structurally attached to the engine. The transmission is attached to the vehicle structure and supports the loads created instead of the engine. In practice, though, many of the same engines are built in both turboprop and turboshaft versions, with only minor differences.

An unusual example of the turboshaft principle is the Pratt & Whitney F135-PW-600 turbofan engine for the STOVL F-35B in conventional mode it operates as a turbofan, but when powering the LiftFan, it switches partially to turboshaft mode to send 29,000 horsepower forward through a shaft and partially to turbofan mode to continue to send thrust to the main engine's fan and rear nozzle.

Large helicopters use two or three turboshaft engines for redundancy. The Mil Mi-26 uses two Lotarev D-136 at 11,400 hp each, while the Sikorsky CH-53E Super Stallion uses three General Electric T64 at 4,380 hp each.

Early turboshaft engines were adaptations of turboprop engines, delivering power through a shaft driven directly from the gas generator shafts, via a reduction gearbox. Examples of direct-drive turboshafts include marinised or industrial Rolls-Royce Dart engines.

RADIAL GAS TURBINES

A radial turbine is a turbine in which the flow of the working fluid is radial to the shaft. The difference between axial and radial turbines consists in the way the fluid flows through the components (compressor and turbine). Whereas for an axial turbine the rotor is 'impacted' by the fluid flow, for a radial turbine, the flow is smoothly orientated perpendicular to the rotation axis, and it drives the turbine in the same way water drives a watermill. The result is less mechanical stress (and less thermal stress, in case of hot working fluids) which enables a radial turbine to be simpler, more robust, and more efficient (in a similar power range) when compared to axial turbines. When it comes to high power ranges (above 5 MW) the radial turbine is no longer competitive (due to heavy and expensive rotor) and the efficiency becomes similar to that of the axial turbines.

Radial turbine.

Advantages and Challenges

Compared to an axial flow turbine, a radial turbine can employ a relatively higher pressure ratio (≈4) per stage with lower flow rates. Thus these machines fall in the lower specific speed and power ranges. For high temperature applications rotor blade cooling in radial stages is not as easy as in axial turbine stages. Variable angle nozzle blades can give higher stage efficiencies in a radial turbine stage even at off-design point operation. In the family of hydro-turbines, Francis turbine is a very well-known IFR turbine which generates much larger power with a relatively large impeller.

Components of Radial Turbines

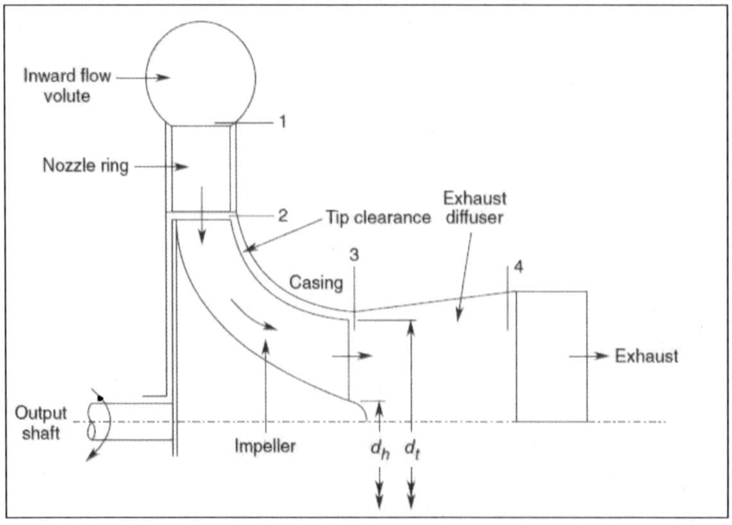

Ninety degree inward-flow radial turbine stage.

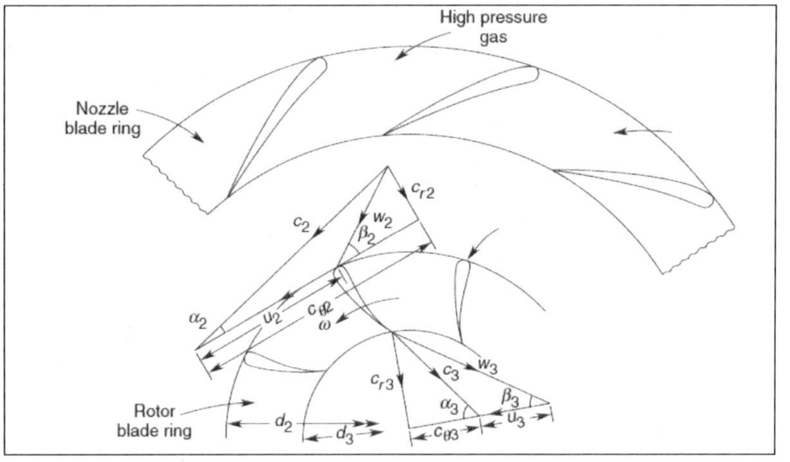

Velocity triangles for an inward-flow radial (IFR) turbine stage with cantilever blades.

The radial and tangential components of the absolute velocity c_2 are c_{r2} and c_{q2}, respectively. The relative velocity of the flow and the peripheral speed of the rotor are w_2 and u_2 respectively. The air angle at the rotor blade entry is given by:

$$\tan \beta_2 = \frac{c_{r2}}{(c_{\theta 2} - u_2)}$$

Enthalpy and Entropy Diagram

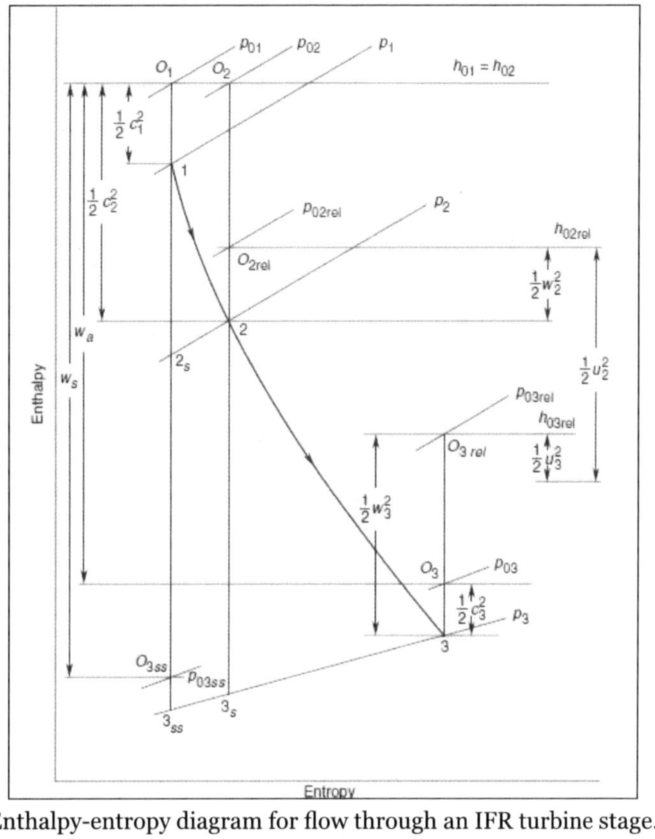

Enthalpy-entropy diagram for flow through an IFR turbine stage.

The stagnation state of the gas at the nozzle entry is represented by point o1. The gas expands adiabatically in the nozzles from a pressure p1 to p2 with an increase in its velocity from c1 to c2. Since this is an energy transformation process, the stagnation enthalpy remains constant but the stagnation pressure decreases (po1 > po2) due to losses. The energy transfer accompanied by an energy transformation process occurs in the rotor.

Spouting Velocity

A reference velocity (c_o) known as the isentropic velocity, spouting velocity or stage terminal velocity is defined as that velocity which will be obtained during an isentropic expansion of the gas between the entry and exit pressures of the stage.

$$C_0 = \sqrt{2C_p T_{01}(1-(\frac{p_3}{p_{01}})^{\frac{\gamma-1}{\gamma}})}$$

Stage Efficiency

The total-to-static efficiency is based on this value of work.

$$\eta_t s = \frac{(h_{01}-h_{03})}{(h_{01}-h_{3ss})}$$

$$\eta_t s = \frac{\psi u_2^2}{[C_p T_{01}(1-(\frac{p_3}{p_{01}})^{\frac{(\gamma-1)}{\gamma}}]}$$

Degree of Reaction

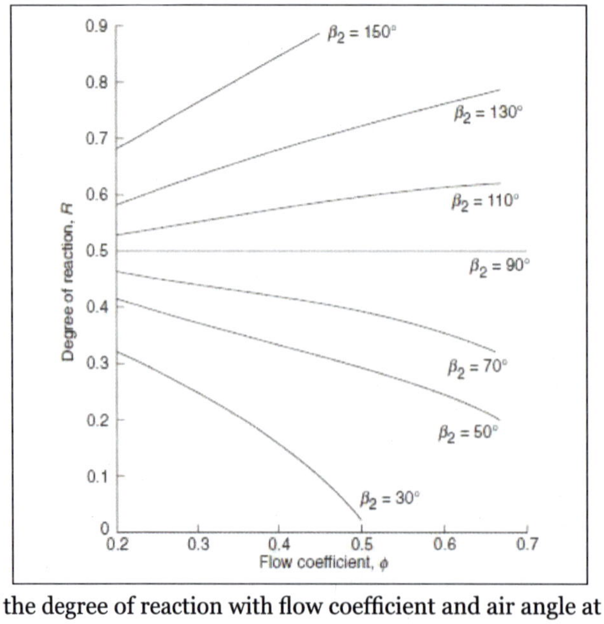

Variation of the degree of reaction with flow coefficient and air angle at rotor entry.

The relative pressure or enthalpy drop in the nozzle and rotor blades are determined by the degree of reaction of the stage. This is defined by:

$$R = \frac{\text{static enthaply drop in rotor}}{\text{stagnation enthalpy drop in stage}}$$

The two quantities within the parentheses in the numerator may have the same or opposite signs. This, besides other factors, would also govern the value of reaction. The stage reaction decreases as $C_{\theta 2}$ increases because this results in a large proportion of the stage enthalpy drop to occur in the nozzle ring.

Stage Losses

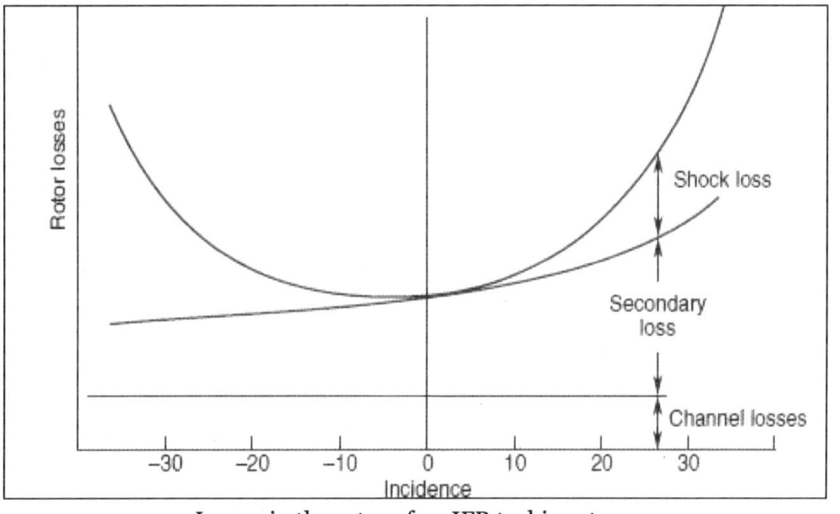

Losses in the rotor of an IFR turbine stage.

The stage work is less than the isentropic stage enthalpy drop on account of aerodynamic losses in the stage. The actual output at the turbine shaft is equal to the stage work minus the losses due to rotor disc and bearing friction.

- Skin friction and separation losses in the scroll and nozzle ring.
  - They depend on the geometry and the coefficient of skin friction of these components.
- Skin friction and separation losses in the rotor blade channels.
  - These losses are also governed by the channel geometry, coefficient of skin friction and the ratio of the relative velocities w_3/w_2. In the ninety degree IFR turbine stage, the losses occurring in the radial and axial sections of the rotor are sometimes separately considered.
- Skin friction and separation losses in the diffuser.
  - These are mainly governed by the geometry of the diffuser and the rate of diffusion.
- Secondary losses.
  - These are due to circulatory flows developing into the various flow passages and are principally governed by the aerodynamic loading of the blades. The main parameters governing these losses are b_2/d_2, d_3/d_2 and hub-tip ratio at the rotor exit.

- Shock or incidence losses.
  - At off-design operation, there are additional losses in the nozzle and rotor blade rings on account of incidence at the leading edges of the blades. This loss is conventionally referred to as shock loss though it has nothing to do with the shock waves.
- Tip clearance loss.
  - This is due to the flow over the rotor blade tips which does not contribute to the energy transfer.

Blade to Gas Speed Ratio

The blade-to-gas speed ratio can be expressed in terms of the isentropic stage terminal velocity c_0.

$$\sigma_s = \frac{u_2}{c_0} = [2(1+\phi_2 \cot \beta_2]^{\frac{-1}{2}}$$

For $\beta_2=90°$ $\sigma_s=0.707$

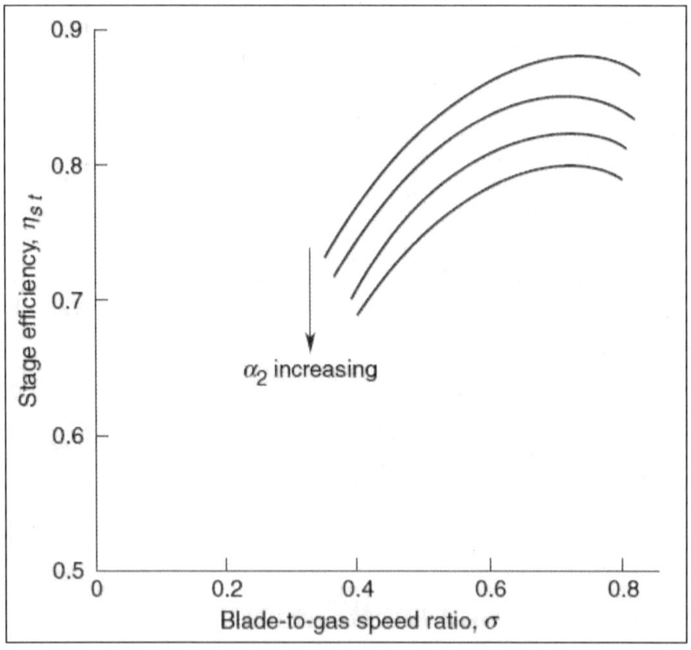

Variation of stage efficiency of an IFR turbine with blade-to-isentropic gas speed ratio.

Outward-flow Radial Stages

In outward flow radial turbine stages, the flow of the gas or steam occurs from smaller to larger diameters. The stage consists of a pair of fixed and moving blades. The increasing area of cross-section at larger diameters accommodates the expanding gas.

This configuration did not become popular with the steam and gas turbines. The only one which is employed more commonly is the Ljungstrom double rotation type turbine. It consists of rings of cantilever blades projecting from two discs rotating in opposite directions. The relative peripheral velocity of blades in two adjacent rows, with respect to each other, is high. This gives a higher value of enthalpy drop per stage.

Nikola Tesla's Bladeless Radial Turbine

In the early 1900s, Nikola Tesla developed and patented his bladeless Tesla turbine. One of the difficulties with bladed turbines is the complex and highly precise requirements for balancing and manufacturing the bladed rotor which has to be very well balanced. The blades are subject to corrosion and cavitation. Tesla attacked this problem by substituting a series of closely spaced disks for the blades of the rotor. The working fluid flows between the disks and transfers its energy to the rotor by means of the boundary layer effect or adhesion and viscosity rather than by impulse or reaction. Tesla stated his turbine could realize incredibly high efficiencies by steam. There has been no documented evidence of Tesla turbines achieving the efficiencies Tesla claimed. They have been found to have low overall efficiencies in the role of a turbine or pump. In recent decades there has been further research into bladeless turbine and development of patented designs that work with corrosive/abrasive and hard to pump material such as ethylene glycol, fly ash, blood, rocks, and even live fish.

MICROTURBINES

Microturbines are small combustion turbines approximately the size of a refrigerator with outputs of 25 kW to 500 kW. They evolved from automotive and truck turbochargers, auxiliary power units (APUs) for airplanes, and small jet engines. Most microturbines are comprised of a compressor, combustor, turbine, alternator, recuperator (a device that captures waste heat to improve the efficiency of the compressor stage), and generator. The figure below illustrates how a microturbine works.

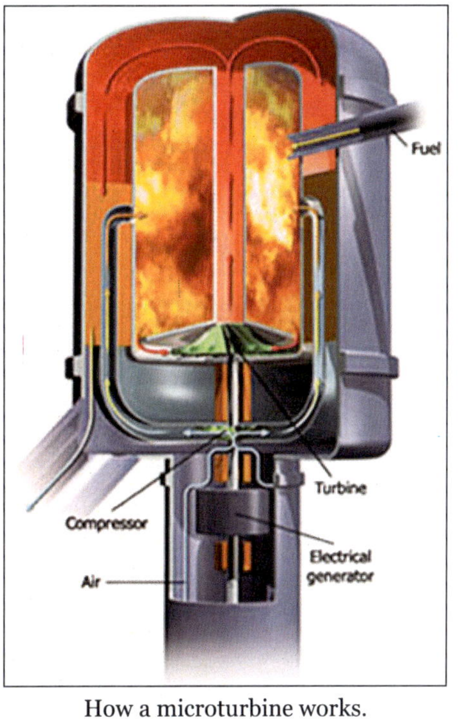

How a microturbine works.

Types of Microturbines

Microturbines are classified by the physical arrangement of the component parts: single shaft or two-shaft, simple cycle, or recuperated, inter-cooled, and reheat. The machines generally rotate over 40,000 revolutions per minute. The bearing selection—oil or air—is dependent on usage. A single shaft microturbine with high rotating speeds of 90,000 to 120,000 revolutions per minute is the more common design, as it is simpler and less expensive to build. Conversely, the split shaft is necessary for machine drive applications, which does not require an inverter to change the frequency of the AC power.

Microturbine generators can also be divided into two general classes:

- Unrecuperated (or simple cycle) microturbines—In a simple cycle, or unrecuperated, turbine, compressed air is mixed with fuel and burned under constant pressure conditions. The resulting hot gas is allowed to expand through a turbine to perform work. Simple cycle microturbines have lower efficiencies at around 15%, but also lower capital costs, higher reliability, and more heat available for cogeneration applications than recuperated units.
- Recuperated microturbines—Recuperated units use a sheet-metal heat exchanger that recovers some of the heat from an exhaust stream and transfers it to the incoming air stream, boosting the temperature of the air stream supplied to the combustor. Further exhaust heat recovery can be used in a cogeneration configuration. The figures below illustrate a recuperated microturbine system. The fuel-energy-to-electrical-conversion efficiencies are in the range of 20 to 30%. In addition, recuperated units can produce 30 to 40% fuel savings from preheating.

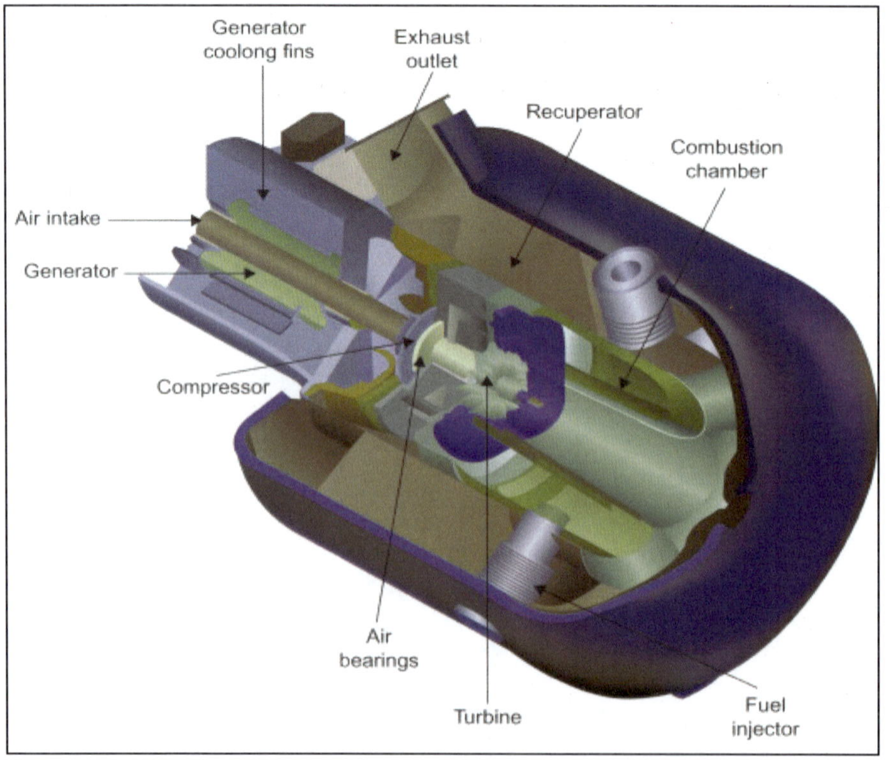

Cogeneration is an option in many cases as a microturbine is located at the point-of-power utilization. The combined thermal electrical efficiency of microturbines in such cogeneration applications can reach as high as 85% depending on the heat process requirements.

Microturbine Efficiency	
Configuration	Efficiency
Unrecuperated	15%
Recuperated	20–30%
With Heat Recovery	Up to 85%

Advanced materials, such as ceramics and thermal barrier coatings, are some of the key enabling technologies to further improve microturbines. Efficiency gains can be achieved with materials like ceramics, which allow a significant increase in engine operating temperature.

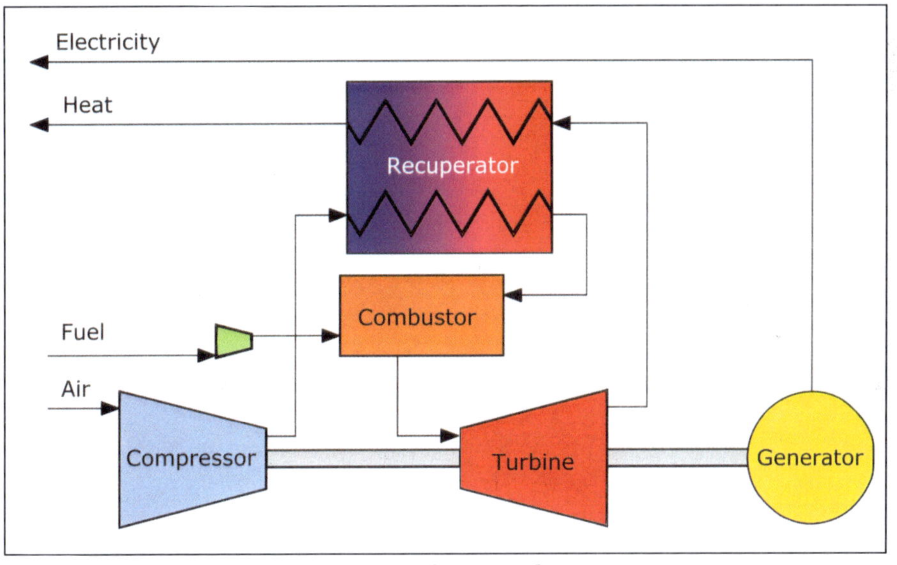

Recuperated microturbine.

Characteristics of Microturbines

Some of the primary applications for microturbines include:

- Distributed generation—Stand-alone, on-site applications remote from power grids.
- Quality power and reliability—Reduced frequency variations, voltage transients, surges, dips, or other disruptions.
- Stand-by power—Used in the event of an outage, as a back-up to the electric grid.
- Peak shaving—The use of microturbines during times when electric use and demand charges are high.
- Boost power—Boost localized generation capacity and on more remote grids.
- Low-cost energy—The use of microturbines as base load or primary power that is less expensive to produce locally than it is to purchase from the electric utility.

- Combined heat and power (cogeneration)—Increases the efficiency of on-site power generation by using the waste heat for existing thermal process.

Microturbines offer many potential advantages for distributed power generation. Selected strengths and weaknesses of microturbine technology are listed in the following table from the California Distributed Energy Resources Guide on Microturbines.

Microturbines	
Strengths	Weaknesses
Small number of moving parts	Low fuel to electricity efficiencies
Compact size	Loss of power output and efficiency with higher ambient temperatures and elevation
Lightweight	
Good efficiencies in cogeneration	
Low emissions	
Can utilize waste fuels	
Long maintenance intervals	
No vibrations	
Less noise than reciprocating engines	
Strengthens energy security	

Accessible: N/A

Aesthetics

- Improves sightlines and views with off-the-grid systems, which eliminate the need for overhead power lines.

Cost-Effective

- Enables cost savings by reducing the peak demand at a facility, therefore lowering demand charges.

Functional

- Provides better power reliability and quality, especially for those in areas where brown-outs, surges, etc. are common or utility power is less dependable.
- Provides power to remote applications where traditional transmission and distribution lines are not an option such as construction sites and offshore facilities.
- Can be an alternative to diesel generators for on-site power for mission critical functions (e.g., communications centers).
- Possesses combined heat and power capabilities.
- Reduces upstream overload of transmission lines.

- Optimizes utilization of existing grid assets—including potential to free up transmission assets for increased wheeling capacity.
- Improves grid reliability.
- Facilitates faster permitting than transmission line upgrades.
- Can be located on sites with space limitations for the production of power.

Productive

- Provides high-quality power for sensitive applications.
- Responds faster to new power demands—as capacity additions can be made more quickly.
- Facilitates less capital tied up in unproductive assets—as the modular nature of microturbines means capacity additions and reductions can be made in small increments, closely matched with demand, instead of constructing central power plants sized to meet estimated future (rather than current) demand.
- Stand-by power decreases downtime, enabling employees to resume working.
- Produces less noise than reciprocating engines.

Secure/Safe

- Strengthens energy security.
- Stand-by power provides quick recovery after an event.

Sustainable

- Produces the lowest emission of any noncatalyzed fossil fuel combustion system.
- Has a small footprint, minimizing site disturbance.
- Reduces or defers infrastructure (line and substation) upgrades.
- For recuperated microturbine, possesses higher energy conversion efficiencies than central generation.
- Enables more effective energy and load management.

Economics of Microturbines

Microturbine capital costs range from $700–$1,100/kW. These costs include all hardware, associated manuals, software, and initial training. Adding heat recovery increases the cost by $75–$350/kW. Installation costs vary significantly by location but generally add 30-50% to the total installed cost.

Microturbine manufacturers are targeting a future cost below $650/kW. This appears to be feasible if the market expands and sales volumes increase.

With fewer moving parts, microturbine vendors hope the units can provide higher reliability than conventional reciprocating generating technologies. Manufacturers expect that initial units will require more unexpected visits, but as the products mature, a once-a-year maintenance schedule should suffice. Most manufacturers are targeting maintenance intervals of 5,000-8,000 hours.

Maintenance costs for microturbine units are still based on forecasts with minimal real-life situations. Estimates range from $0.005–$0.016 per kWh, which would be comparable to that for small reciprocating engine systems.

Microturbine Cost	
Capital Cost	$700–$1,100/kW
O&M Cost	$0.005-0.016/kW
Maintenance Interval	5,000-8,000 hrs

Application

Microturbines can be used for stand-by power, power quality and reliability, peak shaving, and cogeneration applications. In addition, because microturbines are being developed to utilize a variety of fuels, they are being used for resource recovery and landfill gas applications. Microturbines are well suited for small commercial building establishments such as: restaurants, hotels/motels, small offices, retail stores, and many others.

The variety of energy consumers that are already using microturbines is large and growing fast. For example:

- Landfill gas-fired microturbines installed at the Jamacha Landfill in Spring Valley, California supply power on-site and back to the grid.
- A McDonald's restaurant in Chicago, Illinois, gets most of its electricity from a natural-gas-powered microturbine, cutting $1,500 off its total monthly power bill.

A Parallon 75 microturbine generator power system at the Energy Efficient McDonald's (TEEM) in Bensenville, Illinois, is able to power the entire store, including lighting, cooking equipment, and HVAC.

- A textile mill in Lawrence, Massachusetts, ensures continuous operation by getting its power from microturbines.
- The Chesapeake Building on the University of Maryland campus, College Park, Maryland has a cooling, heating, and power (CHP) system consisting of microturbines, chiller, and stack that uses waste heat to cool and heat the building, significantly increasing system efficiency.

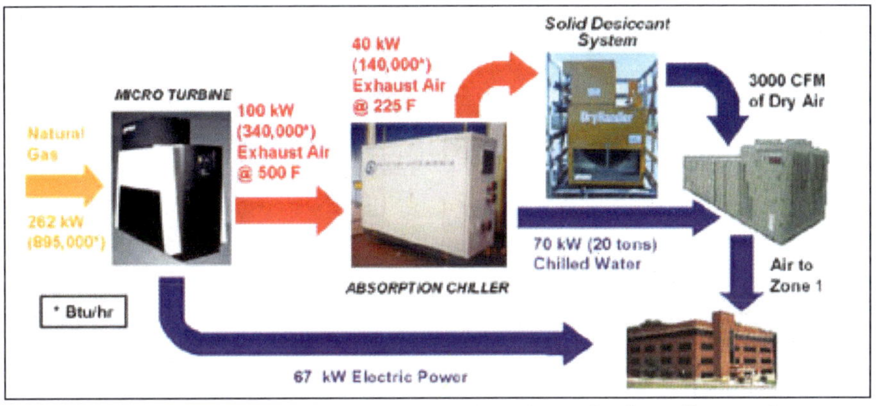

Chesapeake Building CHP system, University of Maryland—College Park, MD.

The development of microturbine technology for transportation applications is also in progress. Automotive companies are interested in microturbines as a lightweight and efficient fossil-fuel-based energy source for hybrid electric vehicles, especially buses.

Other ongoing developments to improve microturbine design, lower costs, and increase performance in order to produce a competitive distributed generation product include heat recovery/cogeneration, fuel flexibility, and hybrid systems (e.g., fuel cell/microturbine, flywheel/microturbine).

References

- Jet-engine, technology: britannica.com, Retrieved 18 May, 2019
- How-an-auxiliary-power-unit-works: flyingmag.com, Retrieved 12 March, 2019
- Case-Gas-Turbines, wp-content: americanenergyinnovation.org, Retrieved 04 June, 2019
- 'Turbines, Compressors and Fans 4th Edition' [Author: S M Yahya; publisher: TATA McGraw-Hill Education (2010)] ISBN 9780070707023
- Microturbines, resources: wbdg.org, Retrieved 05 February, 2019
- Kamps, Thomas (2005). Model Jet Engines. Traplet Publications. ISBN 978-1-900371-91-9

Engine Control and Noise Reduction

3 CHAPTER

Gas turbine engine control is one of the main components of gas turbine as it assures operational safety and reliability in aircraft engines. Some of the noise reduction techniques in gas turbines are scarf inlets, forward swept fans, soft stator vanes, chevron nozzles, etc. This chapter has been carefully written to provide an easy understanding of the gas turbine engine control and noise reduction techniques.

GAS TURBINE ENGINE CONTROL DESIGN

Gas turbine engines are constituted of a complex system. Their desired performance can guarantee the aircraft flight safety. This performance is impressed by some engine input controlling functions which are changing with the development of engines. Finding these functions can be a great success in jet engine control issue. For example, when an airplane loses one of its operative engines during take off, the remaining engines should achieve their maximum thrust as fast as possible without any spools over speed, turbine inlet over temperature, and compressor surge. On the other hand, at a complex maneuvering by a fighter airplane, the engines should deliver more accurate thrust functions. Both of these examples can indicate the importance of the engine control issue.

While the traditional control techniques for aero-engines are time-tested and reliable, modern control techniques promise to provide improved control and therefore improved aircraft propulsion system performance.

During the last five decades, most of control problems have been formulated by the objective knowledge of the given systems (e.g., mathematical model). Many of these "model-base" approaches have found their way into practice and provided satisfactory solutions to the spectrum of complex systems.

However, there are many knowledge-based systems that cannot be merely described by the traditional mathematical representations.

The knowledge-based approaches are much closer to human thinking than traditional classical approaches.

In a broad perspective, knowledge-based approaches underlie what is called "soft computing". These methods include fuzzy logic (FL), neural networks (NN), genetic algorithms (GA), and probabilistic reasoning (PR). In addition, these methodologies in most part are complimentary rather than competitive.

Fuzzy logic is one of the most effective approaches for intelligent control of complex nonlinear systems like aero-engines. One of the important advantages of this approach is the simplicity of utilization. The other one is feasibility of increasing the number and type of membership functions and rules while it has wide variety of rules definition.

The engines of today are reliable and safe, but they are expensive to operate and maintain. With the integration of available models and algorithms for on-board operation and the resultant increase in on-board intelligence, engines of the future will be able to operate even more safely and reliably with reduced life cycle cost.

With a renewed emphasis on reducing engine life cycle costs, improving fuel efficiency, increasing durability and life, and so forth, driven by various development programs, there is a strong push to do research on application of intelligent technologies for the engines.

The intelligent controlling approaches will provide the required scope for aero-engines to be more efficient, safe, and economic. These approaches offer the potential for creating extremely safe, highly reliable systems. The approaches will help to enable a level of performance that far exceeds that of today's propulsion systems in terms of reduction of harmful emissions, maximization of fuel efficiency, and minimization of noise, while improving system affordability and safety.

During the last decade, some research works have been carried out for intelligent control of gas turbine engines using fuzzy logic. In, the authors studied a combination of two potential techniques; fuzzy logic and evolutionary algorithms for a specific gas turbine. The controlling parameters include Inlet Guide Vane (IGV) angle and nozzle area. Similar work which has been presented in is designing and evaluating two types of FLCs by controlling the combustor pressure. However, in the foresaid works and other related studies, on the one hand, a comprehensive FLC has not been proposed that takes the flame-out and the safety limitations into account. Or on the other hand, the effect of fuel flow rate is not considered as a remarkable controlling parameter.

Generally, designing an appropriate controller for each specific aero-engine includes two steps: first, an accurate aero-engine model to simulate the engine behavior individually and second, system recognition. The system recognition as the most important step in creating suitable membership functions and rules is the knowledge of engine performance parameters and specially their relation during engine operation. One of the most important performance parameters is the fuel flow rate and its relation with turbine inlet temperature and compressor surge margin during various engine acceleration maneuverings.

Here, after designing an accurate simulator model for the engine, we consider the fuel flow rate directly as the most comprehensive controlling parameter which has significant effects on all engine performance parameters. A successful approach for achieving the system recognition is also presented. The proposed approach includes designing and training a neural network which simulate the engine performance backwardly. So it empowered us to obtain a series of fuel flow functions for various engine acceleration maneuverings. These functions were used in the process of membership functions definition.

Turbine Engine Simulator Model (TESM)

Engine performance characteristics at design point are important parameters obtained from engine design process. For initial definition work, the operating condition where an engine will spend most time has been traditionally chosen as the engine design point. Alternatively some important high power condition may be chosen. Either way, at the design point the engine configuration, component design and cycle parameters are optimised. A change to the engine design requires a different engine geometry, at a fixed operating condition.

The objective here is not to design an engine. Hence, there is a specific engine with fixed geometry and known parameters. We only seek to simulate different operating conditions of the engine including steady and transient operations.

There are several ways to do engine modeling and simulation such as traditional approaches, FORTRAN, or other 4th generation languages to calculate thermodynamics cycle parameters associated with the design and performance prediction of gas turbine engines. These cycle decks have been used extensively by developers of gas turbine engines to understand the behavior of jet engine designs prior to, during, and after the development of the physical engines. In addition to them, Simulink software and associated tools can be applied in the development of detailed, physics-based, turbine engine models.

Simulink provides an easy-to-use, graphical, modeling and simulation development environment for developing time-based simulations in a wide range of applications, and Simulink is capable of code generation using associated tools. So, a dynamic model for a specific turbojet engine is developed using the MATLAB simulation environment and its Simulink toolbox. The schematic configuration of the simulated turbojet engine is shown in figure.

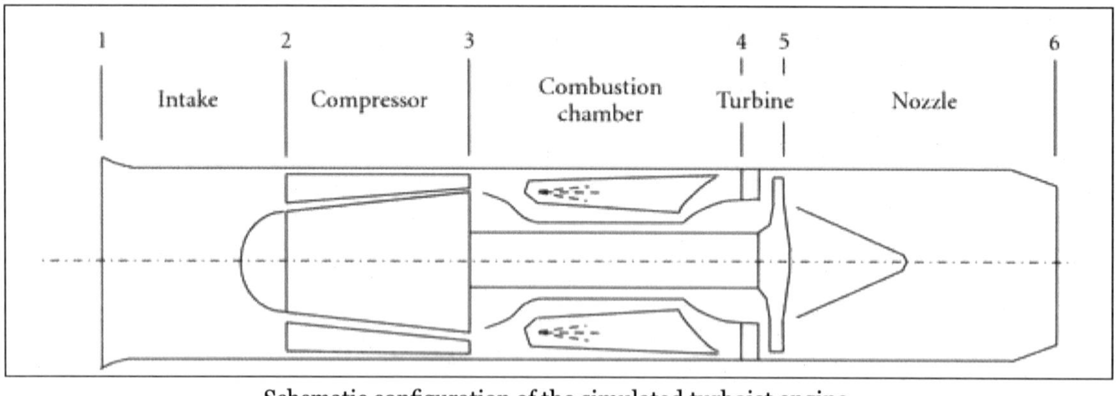

Schematic configuration of the simulated turbojet engine.

The engine model is constructed with a component approach for ease of modification and replacement with different engine components. The precision of Turbine Engine Simulator Model (TESM) results depends on the precision of each component module result.

Each component can be instantiated from a software library module developed to represent the functions of that particular type of component. Each module is a functional unit with its own set of inputs and outputs (I/O). Each can function as an independent component. For example, the intake (Diffuser) module can be used as a stand-alone intake component, and it can be used to instantiate intake in the engine model. The intake module and its I/O are shown in figure.

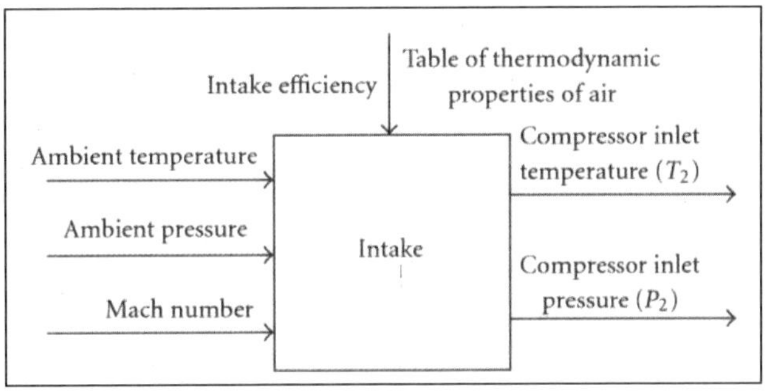

Intake module.

TESM includes some component modules such as: Intake, Compressor, Combustor, Turbine, and Nozzle which are modeled as lumped parameter thermodynamic systems. It means that a multiple stage compressor or turbine is simulated as one component. This approach is adopted because compressor and turbine maps are created to represent the performance of the overall component. The model can be developed to a stage-by-stage fashion if we have required authentic performance data of each component.

The modules are developed based on fundamental laws of physics such as conservation of mass, momentum, and energy. For example, the rotor dynamics is presented by the equation of conservation of angular momentum, that is based on moment of inertia of the components attached to the shaft.

$$\dot{N} = \frac{3600}{4\pi^2 NI}\left[W_4 C_{P4}(T_4 - T_5) - \frac{W_3 C_{P3}(T_3 - T_2)}{\eta_{mech}}\right],$$

where N is engine spool speed, T is temperature, W is mass flow, I is engine spool moment of inertia, and η_{mech} is the engine spool mechanical efficiency.

Other most important dynamical differential equations are used in the modeling process are as follows:

$$\dot{\rho} = \frac{W_3 - W_4 + u}{C_{\text{vol}}},$$

$$\dot{T}_4 = \frac{C_{P3}T_3W_3 - C_{P4}T_4W_4 + u\,\text{LHV}}{C_{\text{vol}}\dot{\rho}_4 T_4 C_{\text{V}}}.$$

In order to use individual components to create an engine model, all hardware component information such as flow, pressure ratio and efficiency maps, flow volume, and area of some components is required.

The incorporation of turbo machinery map data (compressor and turbine maps) into the MATLAB workspace (Simulink-related blocks) is a significant part of the tuning process required to match the model performance with data for a specific engine.

The other parameters to be incorporated into the MATLAB workspace are characteristic lengths, volumes, moments of inertia, design constants, efficiencies, and so forth. Values for these tuning parameters must be determined from engine design documentation, experimentation, or other simulation models. The model references these parameters at run-time.

One of TESM outputs is compressor surge margin which is highly important for safe engine operation. Compressor surge is a condition affecting both aerospace and industrial compression systems that employ turbomachinery. Surge is an unstable operating condition that can lead to the loss of an aircraft in aerospace applications and cause severe damage to industrial systems. During this compression system instability, the flow over the blades of the compressor stalls, and the pressure rise capability is reduced. When this occurs, the compressor cannot maintain the high-pressure downstream and a violent flow reversal occurs throughout the compression system.

Surge Margin (SM) of compressor refers to a margin of safety between the normal operating point of the compressor and the stability limit. Surge Margin is obtained using the incorporated turbo machinery map data, as follows:

$$\mathrm{SM} = \left(\frac{(P_{32} / \dot{m}_{2a_{cor}})_{\mathrm{stall}} - (P_{32} / \dot{m}_{2a_{cor}})}{(P_{32} / \dot{m}_{2a_{cor}})} \right), \quad P_{32} = \left(\frac{P_3}{P_2} \right).$$

The model configuration is shown in figure. As can be seen, the relationships between components are easily understood and the gas path flow parameter connections between major engine components include mass flow rate, total temperature and pressure, static pressure, and Fuel-Air Ratio (FAR).

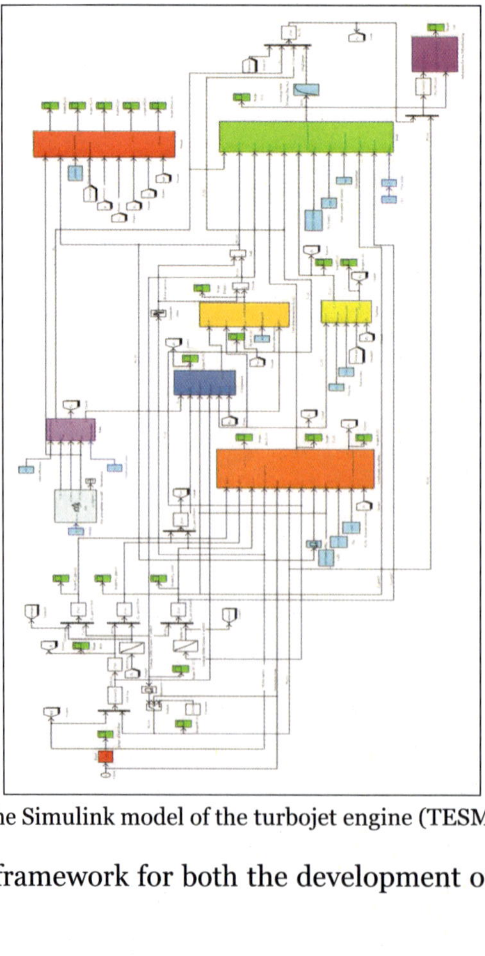

The Simulink model of the turbojet engine (TESM).

This model provides a basic framework for both the development of engine component modules and analysis of component interactions at the system level within the engine. Component modules can be reused and the time and cost required to produce new models significantly reduced.

TESM has three main inputs including altitude and Mach number, which are ambient inputs (Flight Data), and fuel flow to the combustor chamber. The inputs are detailed below.

- Altitude: This is an ambient input with a range of 10,000 feet for this type of engine which goes into the ISA atmosphere model block.
- Mach number: This simulates the speed of aircraft and its effects on P_2, T_2, and C_a The range for this type of engine Mach number is 0 to 0.9.
- Fuel flow: This is an important input from an engine control point of view and is set by a controller for a specific demanded Power Lever Angle (PLA).

The engine transient operation can be indicated by TESM's outputs. Table presents some of these outputs.

Table: Turbine engine simulator model outputs.

Component module	Output parameters
ISA.Atms.Model	a, ρ, P_1, T_1
Intake	P_2, T_2
Compressor	P_3, T_3, W_3
Combustor	P_4, T_4, W_4
Turbine	P_5, T_5
Nozzle	Thrust, SFC, P_{s6}, T_{s6}, C_6
Shaft	N
Surge margin model	SM

Among the model outputs, four are the most important, including: engine spool speed (N), turbine inlet temperature (T_4), compressor surge margin (SM), and engine thrust. These should be controlled by engine controllers. Figure presents TESM and its more important inputs and outputs.

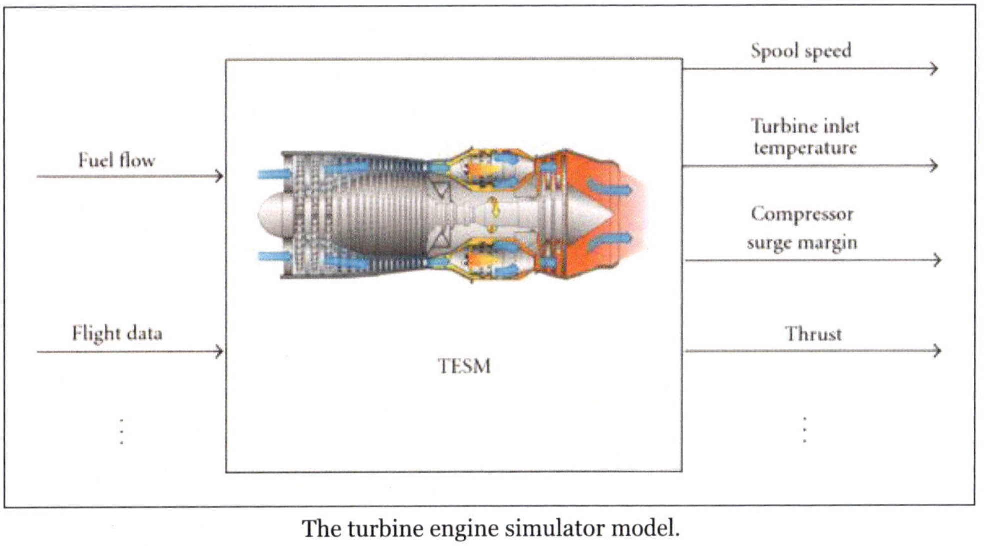

The turbine engine simulator model.

The model operates as a virtual test cell and enables a user to investigate "what-if" scenarios at a fraction of the cost of an engine test cell or research aircraft.

Now, to present an efficient method to estimate the fuel flow injection function to the combustor chamber which is of great importance among the engine input controlling functions, some different reasonable fuel flow functions are applied via the engine model, and then some important engine continuous time operation parameters (such as thrust, compressor surge margin, turbine inlet temperature, and engine spool speed, etc.) are obtained. These parameters and the fuel flow functions provide a precious database which can be used by a neural network.

Each of the foresaid fuel flow functions as an important part of the database is a series of two-dimensional vectors. Since the engine accelerates from normalized spool speed of $n = 0.55$ to $n = 1.1$, the starting and ending points of the functions concern the related steady state values of fuel flow. As a general rule, the Fuel-Air Ratio (FAR) put some constraints on variation of fuel flow rate. In this regard, the derivative of a fuel flow function should be restricted by some prespecified values for the engine. By considering all of the foresaid constraints and the compressor map, a series of discrete points would be obtained. These points are used to estimate the fuel flow functions.

Neural Network Training

Neural networks are a powerful technique to solve many real-world problems. They have the ability to learn from experience in order to improve their performance and to adapt themselves to changes in the environment. In addition to that, they are able to deal with incomplete information or noisy data and can be very effective especially in situations where it is not possible to define the rules or steps that lead to the solution of a problem.

By designing and training a feedforward multilayer perceptron neural network according to the available database; we estimate a number of different reasonable fuel flow functions providing the desired engine performance parameters such as thrust and compressor surge margin for various acceleration maneuverings. The selected engine maneuverings only include twelve fast and safe accelerations for pilot's full throttle commands to cover the critical range of engine operation. The different rates of full throttle commands will diversify these maneuverings. In fact, the proposed neural network simulates the engine performance backwardly. So a series of fuel flow functions would be obtained for various engine acceleration maneuverings. These functions will be used in the process of membership functions definition for designing a fuzzy logic controller.

The network input layer consists of four neurons which are the most important engine performance parameters and acquired by the engine model. These engine model output parameters are depicted in figure. There are ten neurons in the hidden layer. The output layer includes one neuron as the fuel flow rate. The proposed neural network is shown in figure.

We use back propagation algorithm as a reliable and well-known algorithm to train the network. Applying this algorithm would minimize output error by changing the synapse weights.

Three hundred training data and fifty testing data are used by the network. The network activity function is as follows:

$$f(x) = \frac{1}{1+e^{-x}}.$$

Since the above function provides normalized outputs between zero and one; so the inputs of the network should be normalized too. At last, the proposed network would achieve to acceptable error by 3500 epochs and training process finish.

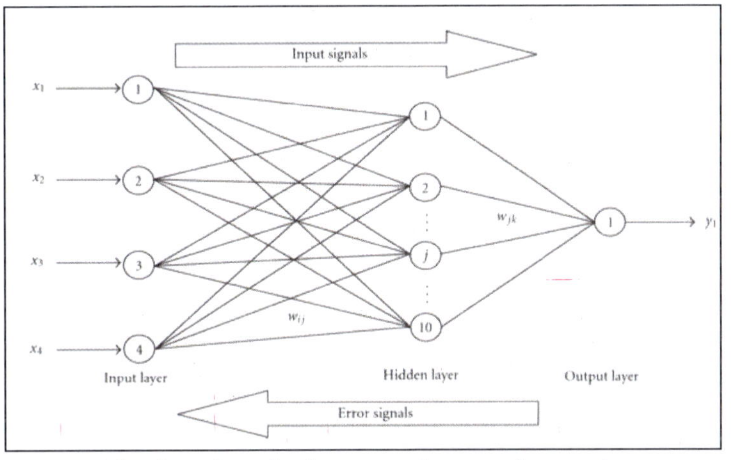

The three-layer back propagation neural network.

Designing a Fuel Flow Controller by the Aid of Fuzzy Logic Method

Fuzzy logic is an increasingly popular method of handling systems associated with uncertainty, unmodeled dynamics, or simply where human experience is required. Its ability to deal with imprecise data can often offer an immediate benefit over conventional mathematical reasoning. It has been widely employed in control problems, particularly due to its ability to mimic the behavior of nonlinear plants. By ensuring that a properly formulated rule base is found, a fuzzy system can provide smooth transitions between operating regimes.

An FLC utilizes fuzzy logic to convert linguistic information based on expert knowledge into an automatic control strategy. In order to use the fuzzy logic for control purposes, a front-end "fuzzifier" and a rear-end "defuzzifier" are added to the usual input-output data set. An FLC commonly consists of four sections: rules, fuzzifier, inference engine, and defuzzifier. Once the rule has been established, the controller can be considered as a nonlinear mapping from the input to the output. The block diagram of the generalized indistinct controller consists of four elements:

- Fuzzification block, transforming input physical values y_i into corresponding linguistic variables μ(yi);
- Fuzzy rule base, including:

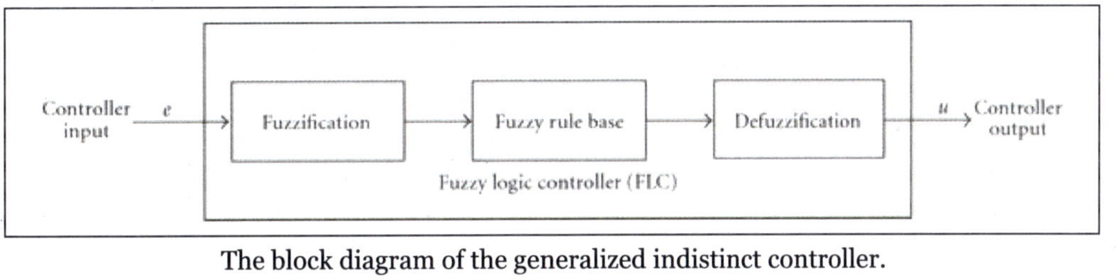

The block diagram of the generalized indistinct controller.

- Knowledge base, containing rules table for logic output block;

- Logic output block, transforming input linguistic variables into output with some belonging functions;
- Defuzzification block, transforming output linguistic variables into physical control influence.

The designed FLC would determine the amount of fuel flow to the combustor chamber over its transient operation. The FLC and the available turbine engine model constitute a closed loop which is shown in figure.

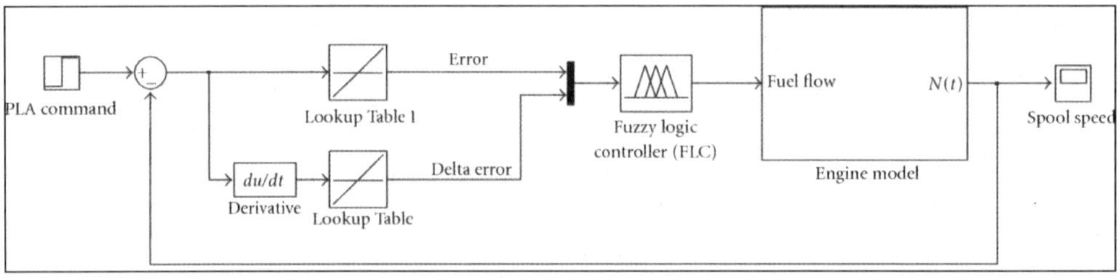

Layout of the turbojet engine simulator model and FLC.

The first input of controller is the spool speed error (the difference of throttle demanded spool speed and current spool speed). This input is named Error. The defined fuzzy function for the input is shown in figure. The second input is the first input variations over time. This input is named Delta Error. The defined fuzzy function for the second input is shown in figure.

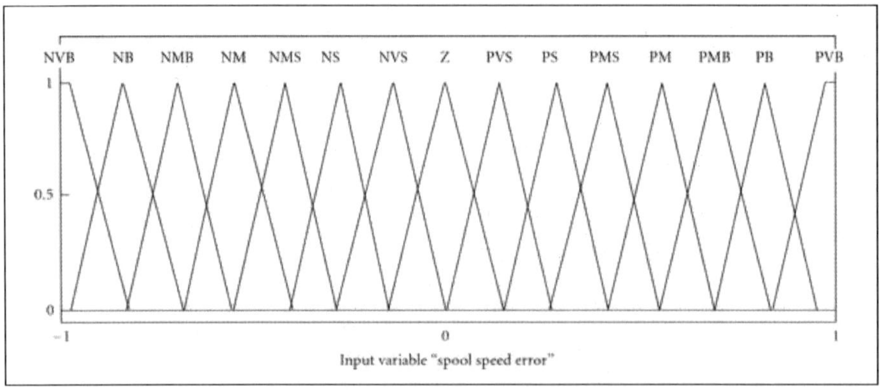

Error membership functions.

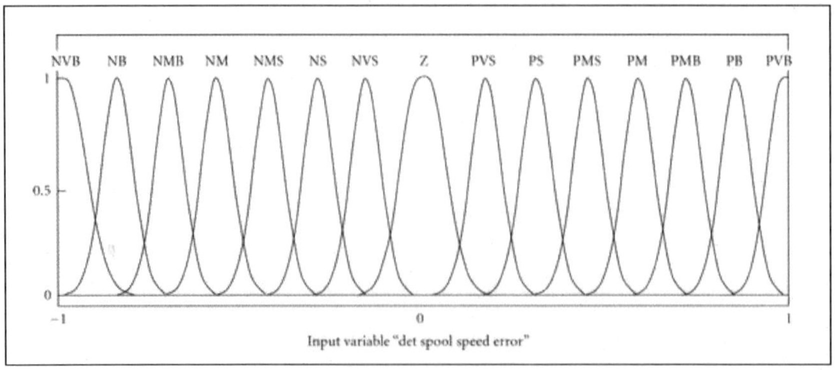

Delta error membership functions.

Defining the proper fuzzy membership functions for fuel flow has a great influence on the controller behavior. So, the obtained desired fuel flow functions as the neural network results would be helpful for defining the fuel flow fuzzy membership functions. This approach is helpful to defuzzify the fuzzy fuel flow membership functions.

In this regard, each of the obtained desired fuel flow functions is separately considered over fifteen equispaced time intervals in [0.0, 3.5]. So an equal number of nonequispaced intervals are also obtained on each fuel flow axis. Since the rate of fuel flow injection is much higher over the starting time intervals (starting phase of acceleration), so a large number fuel flow intervals are situated close to the end part of their axis. Fuel flow intervals on a sample desired fuel flow function are depicted by different arrows in figure.

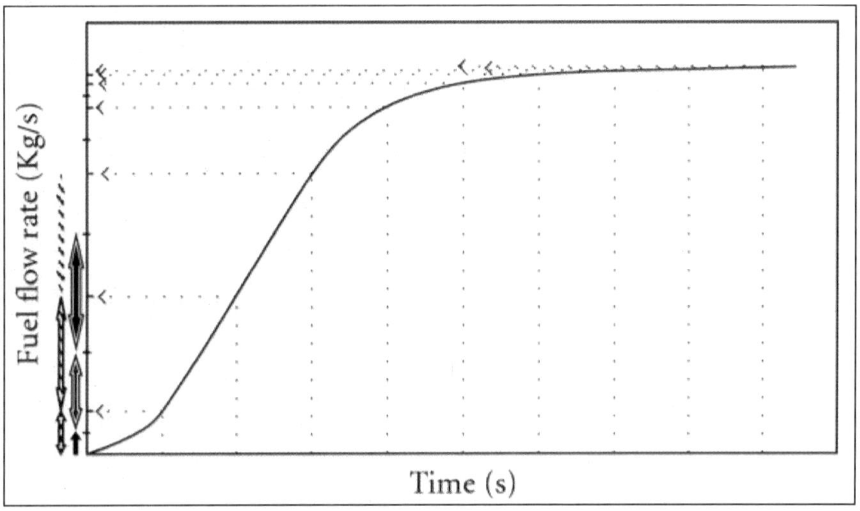

Fuel flow interval construction on a desired sample fuel flow function.

Using a simple mapping method, each of these intervals creates a triangular fuel flow membership function. According to the proposed mapping, the membership degree for midpoint of an interval is one and for endpoints is zero. With respect to the proposed arrangement of intervals in figure, the first three of fuel flow fuzzy membership functions are presented in table.

Table: The first three of fuel flow intervals and the related membership degrees.

	NVB			NB			NMB		
Function name									
Fuel interval points	First	Central	End	First	Central	End	First	Central	End
Fuel amounts	-	0.036	0.041	0.036	0.042	0.047	0.042	0.051	0.062
Membership degree	-	1.0	0.0	0.0	1.0	0.0	0.0	1.0	0.0

A similar approach is applied for each desired fuel flow function. Then, the obtained fuzzy fuel flow functions are used to define a final set of fuel flow fuzzy membership functions. In this regard, for each fuel interval the Arithmetic mean of fuel values with the highest membership degree is obtained and labeled with maximum membership degree. These final membership functions are used in designing procedure of the FLC.

Conventionally, fuzzy rules are established by a combination of knowledge, experience, and observation and may thus not be optimal. Additionally, in spite of efforts to formalize a development standard for fuzzy controllers, fine tuning its performance is still a matter of trial and error.

A variety of membership functions can be defined for inputs and output of the controller. By changing the number and types of the functions and rules, the engine behavior would change.

The acceleration time to achieve pilot demanded operation point is of great importance. This time should be short as much as possible. On the other hand, if the acceleration time be shorter than its allowable limit, compressor surge would be likely. So, it is necessary for the FLC to limit the inordinate engine acceleration.

Here, the applied logic for creation the FLC rules includes higher spool acceleration rates during lower spool speeds and lower spool acceleration rates during higher spool speeds. Some selected rules which produce the central part of control surface are presented in table.

Selected fuzzy rules.

Error	Delta-Error						
	NMS	NS	NVS	Z	PVS	PS	PMS
MNS	NMS	NS	NS	NS	NS	NVS	Z
NS	NMS	NS	NS	NS	NVS	Z	PVS
NVS	NMS	NS	NS	NVS	Z	PVS	PS
Z	NMS	NS	NVS	Z	PVS	PS	PMS
PVS	NMS	NS	NVS	Z	PVS	PS	PMS
PS	NVS	Z	Z	Z	PVS	PS	PMS
PMS	Z	PVS	PS	PS	PS	PS	PMS

The defuzzification process, which takes place after the generation of the fuzzy control signals, is completed using the inference mechanism. The resulting fuzzy set must be converted to a quantity which would be sent to the process regulating valve as a control signal. In this part, the inference results of all activated logic rules are synthesized into crisp output for making a decision. The logic AND has been implemented here with the minimum operator, and the defuzzification method is based on bisector area.

The variation of fuel flow versus the two controller inputs is depicted in figure.

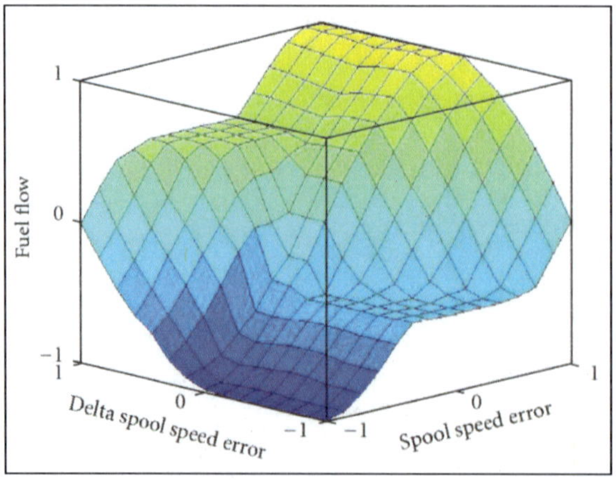

Control surface output of the Fuzzy Logic Controller (FLC).

By designing the controller, the influence of controller parameters on turbine engine should be examined in more detail. So it would be essential to simulate the turbine engine and controller simultaneously.

Testing and Demonstration

The turbine engine acceleration for full throttle command would be considered to achieve the most critical condition. The simulations are run with the unit step power lever angle and constant zero values for Mach number and altitude.

The desirable engine performance would be obtained by applying more and various types of membership functions and rules and then rerunning the simulation program. This technique is effective to achieve the best fuzzy logic controller. The simulation results of engine model with FLC are presented in figure.

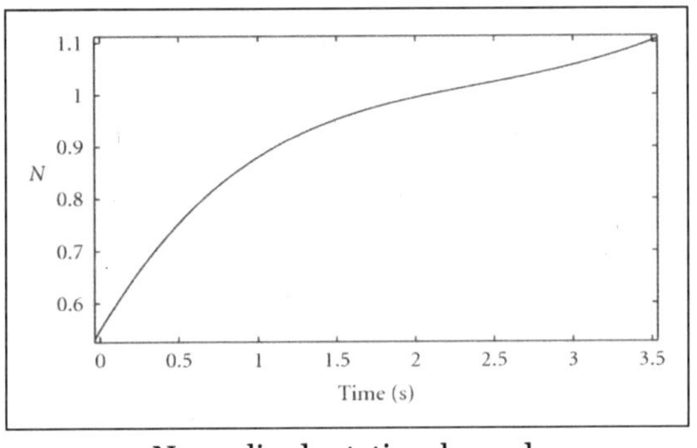

Normalized rotational speed.

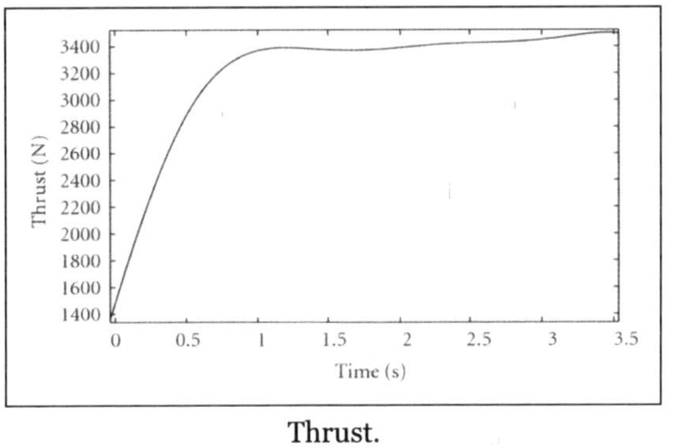

Thrust.

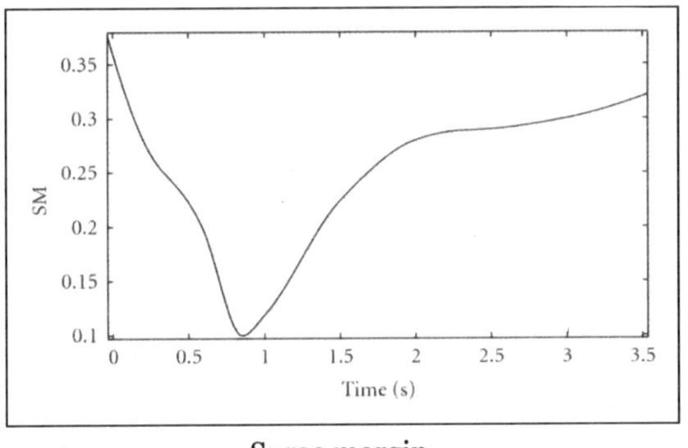

Surge margin.

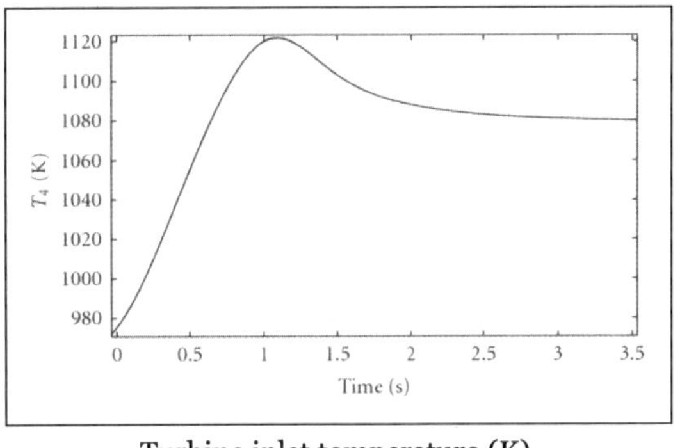

Turbine inlet temperature (K).

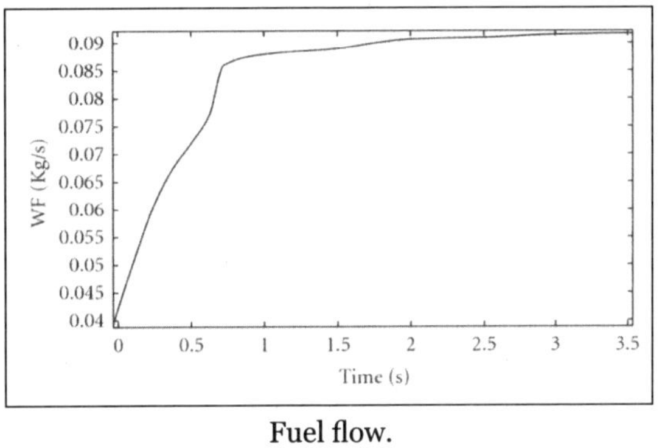

Fuel flow.

When the turbine output power overcomes the compressor input power, the acceleration operation will be done. So, increasing the fuel flow rate is a common method to accelerate the engine spool.

Through the initial time of acceleration, the turbine inlet temperature and the compressor output pressure will be risen and the compressor air flow rate reduced. Considering the said circumstances during engine acceleration, high acceleration rate may lead to compressor surge or stall and excessive turbine inlet temperature which must be controlled by FLC. figures above demonstrate the FLC capability in providing the above requirements.

The acceptability of gas turbine engine is crucially influenced by its ability in producing a demanded thrust while desirable fuel consumption. The demanded thrust for the jet engine is 3500 (N). figure shows the FLC capability in reaching the demanded thrust.

The simulation results of engine model with fuzzy controller in comparison with the engine model with a conventional controller are presented in table.

Table: Simulation results of engine model with FLC in comparison with the engine model with conventional controller.

Parameter	Fuzzy logic controller (simulation)					Conventional controller (simulation)				
Time (Sec)	0	0.8	1.6	2.4	3.2	0	0.8	1.6	2.4	3.2
Spool speed (normalized)	0.55	0.86	0.96	1.02	1.05	0.55	0.74	0.84	0.95	0.98
Turbine inlet temperature (K)	975.39	1118	1098	1086	1081	975.39	1129	1119	1109	1103
Thrust (N)	1482	3404	3462	3471	3478	1482	3379	3393	3409	3412
SM	0.36	0.1	0.23	0.29	0.31	0.36	0.12	0.26	0.33	0.35

As it is seen from table, the proposed FLC has led to more rapid acceleration and lower turbine inlet temperature. However, the SM is reduced in comparison with the conventional controller.

One of important inputs for FLC is the engine spool speed error which is shown in figure. The magnitude of this error varies from its maximum (0.55) at the start of simulation to its minimum (zero) by achieving the demanded spool speed.

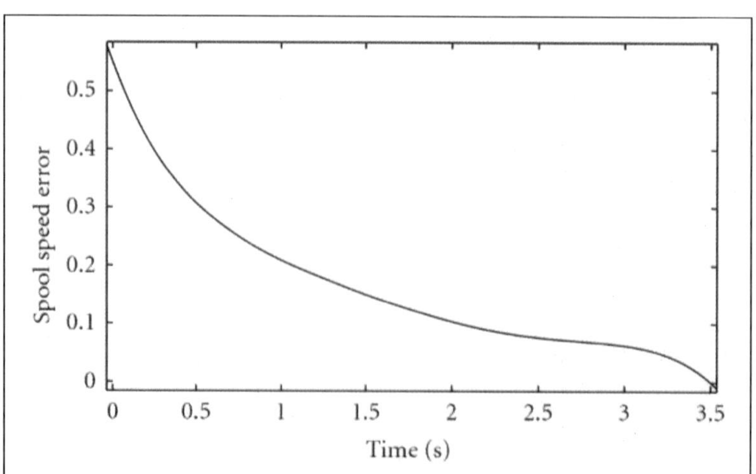

Trace of engine spool speed error.

At the end of specified engine acceleration, some parameters such as time of engine acceleration and thrust are presented in table. The acceleration time has reduced by FLC. This would be an important advantage of the proposed FLC.

Table: Simulation results of engine model with fuzzy controller in comparison with the engine testing operation.

Parameter	Fuzzy logic controller (simulation) $N = 1.1$	Conventional controller (test) $N = 1.1$
Time (sec)	3.5	4.2
Thrust (N)	3484	3420

Whether a fuzzy control design will be stable is a somewhat open question. The problems of FLC stability analysis and optimality are not addressed explicitly; such issues are still open problems in fuzzy controller design. Various nonlinear stability analysis methodologies could be applied for analyzing fuzzy control systems. Some methods have been proposed based on the Lyapunov's second method. The second method of Lyapunov (also referred to as the direct method) is the most general for determining the stability of a nonlinear and/or time-variant system of any order. It is a hot research topic now. In this project, the response curves of the system and its parameters including rise time, overshoot ratio, and settle time may be analyzed for specifying the performance of the FLC. The proper parameters for system response are obtained by adjusting either the rules, the input and output scaling factors, or some other parameters of the FLC. Using the foresaid method, the control loop behavior for the system presented here is well known and considered sufficient.

HYDROMECHANICAL CONTROL SYSTEM

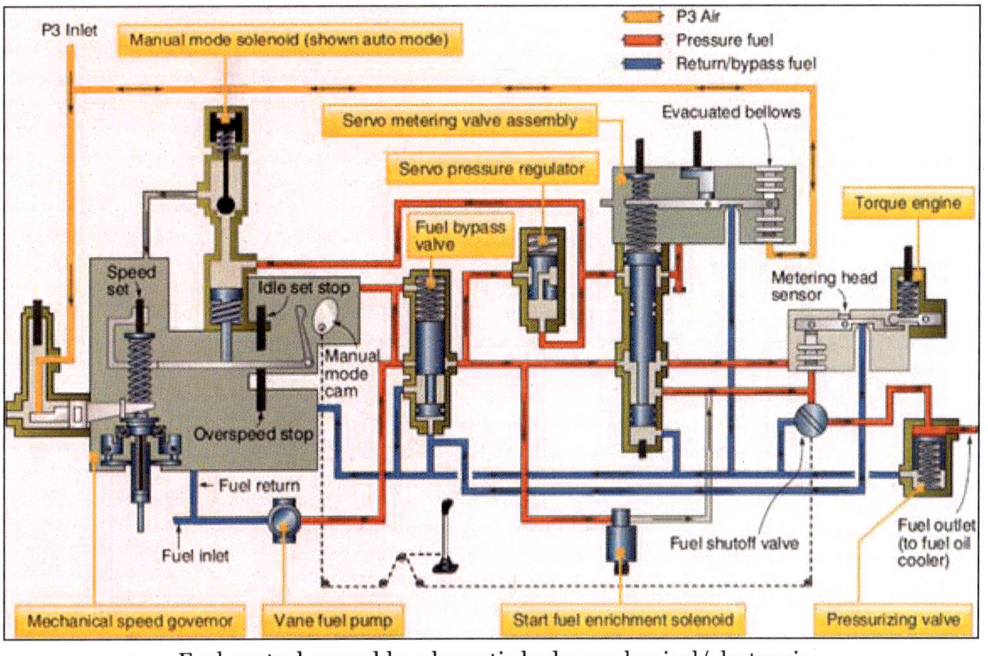

Fuel control assembly schematic hydromechanical/electronic.

Hydromechanical fuel controls were used and are still used on many engines, but their use is becoming limited giving way to electronic based controls. Fuel controls have two sections, computing and metering, to provide the correct fuel flow for the engine. A pure hydromechanical fuel control has no electronic interface assisting in computing or metering the fuel flow. It also is generally driven by the gas generator gear train of the engine to sense engine speed. Other mechanical engine parameters that are sensed are compressor discharge pressure, burner pressure, exhaust temperature, and inlet air temperature and pressure. Once the computing section determines the correct amount of fuel flow, the metering section through cams and servo valves delivers the fuel to the engine fuel system. Actual operating procedures for a hydromechanical fuel control is very complicated and still the fuel metering is not as accurate as with an electronic type of interface or control. Electronic controls can receive more inputs with greater accuracy than hydromechanical controls. Early electronic controls used a hydromechanical control with an electronic system added on the system to fine tune the metering of the fuel. This arrangement also used the hydromechanical system as a backup if the electronic system failed.

Hydromechanical/Electronic Fuel Control

The addition of the electronic control to the basic hydromechanical fuel control was the next step in the development of turbine engine fuel controls. Generally, this type of system used a remotely located EEC to adjust the fuel flow. A description of a typical system is explained in the following information. The basic function of the engine fuel system is to pressurize the fuel, meter fuel flow, and deliver atomized fuel to the combustion section of the engine. Fuel flow is controlled by a hydromechanical fuel control assembly, which contains a fuel shutoff section and a fuel metering section.

This fuel control unit is sometimes mounted on the vane fuel pump assembly. It provides the power lever connection and the fuel shutoff function. The unit provides mechanical overspeed protection for the gas generator spool during normal (automatic mode) engine operation. In automatic mode, the EEC is in control of metering the fuel. In manual mode, the hydromechanical control takes over.

During normal engine operation, a remotely mounted electronic fuel control unit (EFCU) (same as an EEC) performs the functions of thrust setting, speed governing and acceleration, and deceleration limiting through EFCU outputs to the fuel control assembly in response to power lever inputs. In the event of electrical or EFCU failure, or at the option of the pilot, the fuel control assembly functions in manual mode to allow engine operation at reduced power under control of the hydromechanical portion of the controller only.

The total engine fuel and control system consists of the following components and provides the functions as indicated:

- The vane fuel pump assembly is a fixed displacement fuel pump that provides high pressure fuel to the engine fuel control system.
- The filter bypass valve in the fuel pump allows fuel to bypass the fuel filter when the pressure drop across the fuel filter is excessive. An integral differential pressure indicator visually flags an excessive differential pressure condition before bypassing occurs, by extending a pin from the fuel filter bowl. Fuel pump discharge flow in excess of that required by the fuel control assembly is returned from the control to the pump interstage.

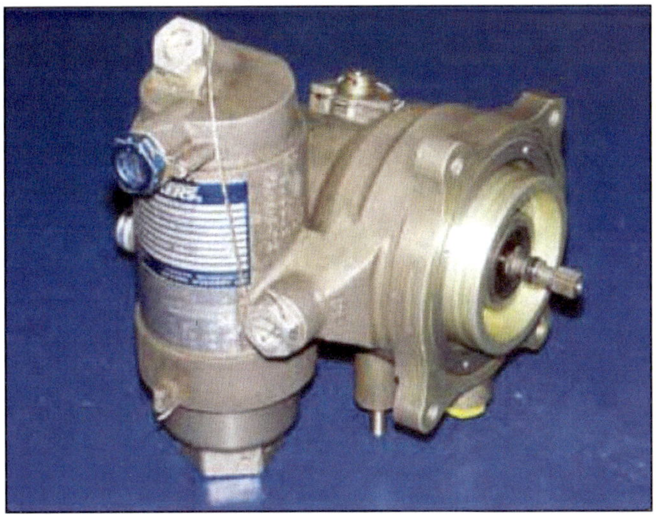

Fuel pump and filter.

- The hydromechanical fuel control assembly provides the fuel metering function of the EFCU.

  Fuel is supplied to the fuel control through a 200-micron inlet filter screen and is metered to the engine by the servo-operated metering valve. It is a fuel flow/compressor discharge pressure (Wf/P3) ratio device that positions the metering valve in response to engine compressor discharge pressure (P3). Fuel pressure differential across the servo valve is maintained by the servo-operated bypass valve in response to commands from the EFCU. The manual mode solenoid valve is energized in the automatic mode. The automatic mode restricts operation of the mechanical speed governor. It is restricted to a single overspeed governor setting above the speed range controlled electronically. Deenergizing the manual mode valve enables the mechanical speed governor to function as an all speed governor in response to power lever angle (PLA). The fuel control system includes a low power sensitive torque motor which may be activated to increase or decrease fuel flow in the automatic mode (EFCU mode). The torque motor provides an interface to an electronic control unit that senses various engine and ambient parameters and activates the torque motor to meter fuel flow accordingly. This torque motor provides electromechanical conversion of an electrical signal from the EFCU. The torque motor current is zero in the manual mode, which establishes a fixed Wf/P3 ratio.

  This fixed wf/P3 ratio is such that the engine operates surge free and is capable of producing a minimum of 90 percent thrust up to 30,000 feet for this example system. All speed governing of the high-pressure spool (gas generator) is achieved by the flyweight governor. The flyweight governor modulates a pneumatic servo, consistent with the speed set point as determined by the power lever angle (PLA) setting. The pneumatic servo accomplishes Wf/P3 ratio modulation to govern the gas generator speed by bleeding down the P3 acting on the metering valve servo. The P3 limiter valve bleeds down the P3 pressure acting in the metering valve servo when engine structural limits are encountered in either control mode. The start fuel enrichment solenoid valve provides additional fuel flow in parallel with the metering valve when required for engine cold starting or altitude restarts. The valve is energized by the EFCU when enrichment is required. It is always deenergized in the manual mode to prevent high altitude sub-idle operation.

Located downstream of the metering valve are the manual shutoff and pressurizing valves. The shutoff valve is a rotary unit connected to the power lever. It allows the pilot to direct fuel to the engine manually. The pressurizing valve acts as a discharge restrictor to the hydromechanical control. It functions to maintain minimum operating pressures throughout the control. The pressurizing valve also provides a positive leaktight fuel shutoff to the engine fuel nozzles when the manual valve is closed.

- The flow divider and drain valve assembly proportions fuel to the engine primary and secondary fuel nozzles. It drains the nozzles and manifolds at engine shutdown. It also incorporates an integral solenoid for modifying the fuel flow for cold-starting conditions.

  During an engine start, the flow divider directs all flow through the primary nozzles. After start, as the engine fuel demand increases, the flow divider valve opens to allow the secondary nozzles to function. During all steady-state engine operation, both primary and secondary nozzles are flowing fuel. A 74-micron, self-bypassing screen is located under the fuel inlet fitting and provides last chance filtration of the fuel prior to the fuel nozzles.

- The fuel manifold assembly is a matched set consisting of both primary and secondary manifolds and the fuel nozzle assemblies.

Twelve fuel nozzles direct primary and secondary fuel through the nozzles causing the fuel to swirl and form a finely atomized spray. The manifold assembly provides fuel routing and atomizing to ensure proper combustion.

The EEC system consists of the hydromechanical fuel control, EFCU, and aircraft mounted power lever angle potentiometer. Aircraft-generated control signals include inlet pressure, airstream differential pressure, and inlet temperature plus pilot selection of either manual or auto mode for the EFCU operation. Engine-generated control signals include fan spool speed, gas generator spool speed, inner turbine temperature, fan discharge temperature, and compressor discharge pressure. Aircraft- and engine-generated control signals are directed to the EFCU where these signals are interpreted. The PLA potentiometer is aircraft mounted in the throttle quadrant. The PLA potentiometer transmits an electrical signal to the EFCU, which represents engine thrust demand in relation to throttle position. If the EFCU determines a power change is required, it commands the torque motor to modulate differential pressure at the head sensor. This change in differential pressure causes the metering valve to move, varying fuel flow to the engine as required. The EFCU receives electrical signals which represent engine operating variables. It also receives a pilot-initiated signal (by power-lever position) representing engine thrust demand. The EFCU computes electrical output signals for use by the engine fuel control for scheduling engine operation within predetermined limits. The EFCU is programmed to recognize predetermined engine operating limits and to compute output signals such that these operating limits are not exceeded. The EFCU is remotely located and airframe mounted. An interface between the EFCU and aircraft/engine is provided through the branched wiring harness assembly.

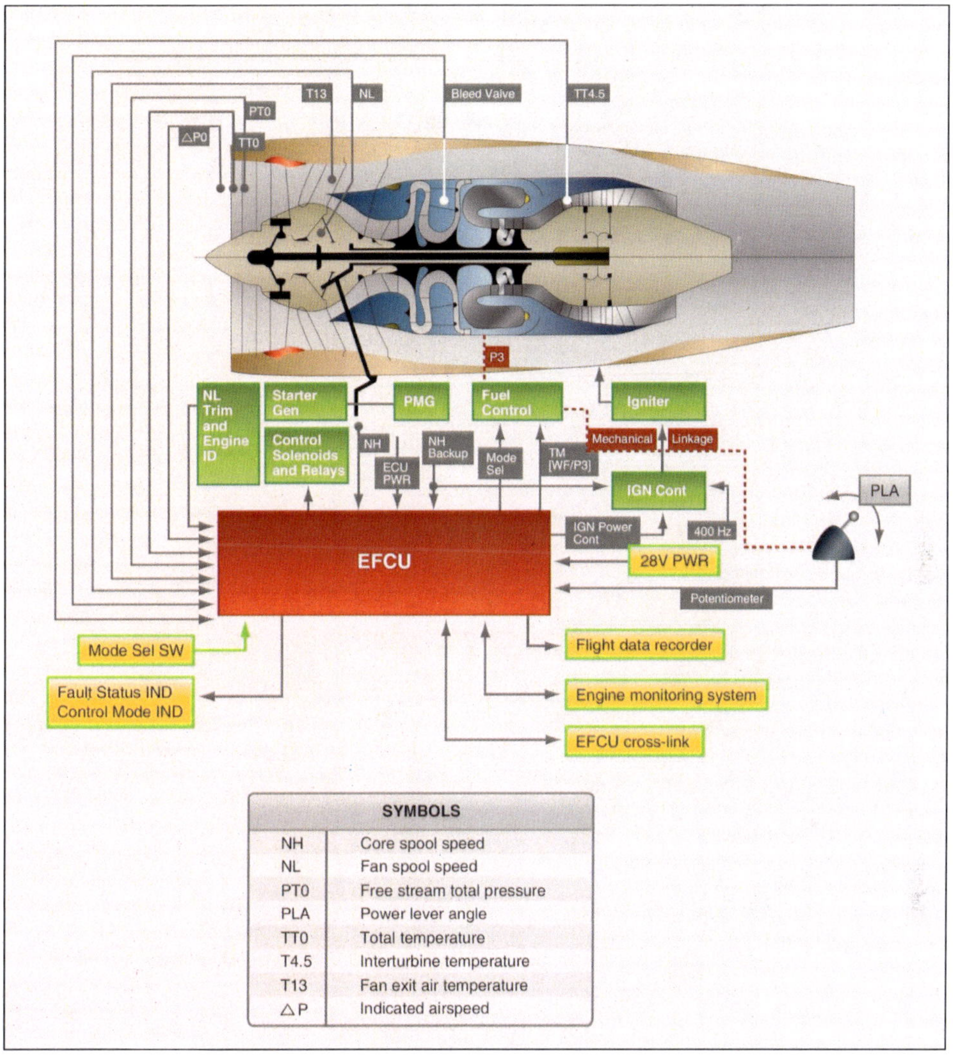

Engine control system.

GAS TURBINE EMISSIONS AND CONTROL

Typical exhaust emissions from a stationary gas turbine. There are two distinct categories. The major species (CO_2, N_2, H_2O, and O_2) are present in percent concentrations. The minor species (or pollutants) such as CO, UHC, NO_x, SO_x, and particulates are present in parts per million concentrations. In general, given the fuel composition and machine operating conditions, the major species compositions can be calculated. The minor species, with the exception of total sulfur oxides, cannot. Characterization of the pollutants requires careful measurement and semitheoretical analysis.

The pollutants shown in table are a function of gas turbine operating conditions and fuel composition. In the following sections, each pollutant will be considered as a function of operating conditions under the broad divisions of gaseous and liquid fuels.

Table: Gas turbine exhaust emissions burning conventional fuels.

Major Species	Typical Concentration (% Volume)	Source
Nitrogen (N_2)	66 - 72	Inlet Air
Oxygen (O_2)	12 - 18	Inlet Air
Carbon Dioxide (CO_2)	1 - 5	Oxidation of Fuel Carbon
Water Vapor (H_2O)	1 - 5	Oxidation of Fuel Hydrogen
Minor Species Pollutants	Typical Concentration (PPMV)	Source
Nitric Oxide (NO)	20 - 220	Oxidation of Atmosphere Nitrogen
Nitrogen Dioxide (NO_2)	2 - 20	Oxidation of Fuel-Bound Organic Nitrogen
Carbon Monoxide (CO)	5 - 330	Incomplete Oxidation of Fuel Carbon
Sulfur Dioxide (SO_2)	Trace - 100	Oxidation of Fuel-Bound Organic Sulfur
Sulfur Trioxide (SO_3)	Trace - 4	Oxidation of Fuel-Bound Organic Sulfur
Unburned Hydrocarbons (UHC)	5 - 300	Incomplete Oxidation of Fuel or Intermediates
Particulate Matter Smoke	Trace - 25	Inlet Ingestion, Fuel Ash, Hot-Gas-Path
		Attrition, Incomplete Oxidation of Fuel or
		Intermediates

Nitrogen Oxides

Nitrogen oxides ($NO_x = NO + NO_2$) must be divided into two classes according to their mechanism of formation. Nitrogen oxides formed from the oxidation of the free nitrogen in the combustion air or fuel are called "thermal NO_x." They are mainly a function of the stoichiometric adiabatic flame temperature of the fuel, which is the temperature reached by burning a theoretically correct mixture of fuel and air in an insulated vessel.

The following is the relationship between combustor operating conditions and thermal NO_x production:

- NOx increases strongly with fuel-to-air ratio or with firing temperature.
- NOx increases exponentially with combustor inlet air temperature.
- NOx increases with the square root of the combustor inlet pressure.
- NOx increases with increasing residence time in the flame zone.
- NOx decreases exponentially with increasing water or steam injection or increasing specific humidity.

Emissions which are due to oxidation of organically bound nitrogen in the fuel—fuel-bound nitrogen (FBN)—are called "organic NO_x." Only a few parts per million of the available free nitrogen (almost all from air) are oxidized to form nitrogen oxide, but the oxidation of FBN to NOx is very efficient. For conventional GE combustion systems, the efficiency of conversion of FBN into nitrogen oxide is 100% at low FBN contents. At higher levels of FBN, the conversion efficiency decreases.

Organic NO_x formation is less well understood than thermal NO_x formation. It is important to note that the reduction of flame temperatures to abate thermal NO_x has little effect on organic NO_x. For liquid fuels, water and steam injection actually increases organic NO_x yields. Organic NO_x formation is also affected by turbine firing temperature. The contribution of organic NO_x is important only for fuels that contain significant amounts of FBN such as crude or residual oils. Emissions from these fuels are handled on a case-by-case basis.

Gaseous fuels are generally classified according to their volumetric heating value. This value is useful in computing flow rates needed for a given heat input, as well as sizing fuel nozzles, combustion chambers, and the like. However, the stoichiometric adiabatic flame temperature is a more important parameter for characterizing NO_x emission. Table shows relative thermal NO_x production for the same combustor burning different types of fuel. This table shows the NO_x relative to the methane NO_x based on adiabatic stoichiometric flame temperature. The gas turbine is controlled to approximate constant firing temperature and the products of combustion for different fuels affect the reported NO_x correction factors. Therefore, Table also shows columns for relative NO_x values calculated for different fuels for the same combustor and constant firing temperature relative to the NO_x for methane.

Typical NO_x performance of the MS7001EA, MS6001B, MS5001P, and MS5001R gas turbines burning natural gas fuel and No. 2 distillate is a function of firing temperature. The levels of emissions for No. 2 distillate oil are a very nearly constant fraction of those for natural gas over the operating range of turbine inlet temperatures. For any given model of GE heavy-duty gas turbine, NO_x correlates very well with firing temperature.

Low-Btu gases generally have flame temperatures below 3500 °F/1927 °C and correspondingly lower thermal NO_x production. However, depending upon the fuel-gas clean-up train, these gases may contain significant quantities of ammonia. This ammonia acts as FBN and will be oxidized to NO_x in a conventional diffusion combustion system. NO_x control measures such as water injection or steam injection will have little or no effect on these organic NO_x emissions.

Carbon Monoxide

Carbon monoxide (CO) emissions from a conventional GE gas turbine combustion system are less than 10 ppmvd (parts per million by volume dry) at all but very low loads for steadystate operation. During ignition and acceleration, there may be transient emission levels higher than those presented here. Because of the very short loading sequence of gas turbines, these levels make a negligible contribution to the integrated emissions. Figure shows typical CO emissions from a MS7001EA, plotted versus firing temperature. As firing temperature is reduced below about 1500°F/816°C the carbon monoxide emissions increase quickly. This characteristic curve is typical of all heavy-duty machine series.

Table: Relative thermal NOx emissions.

Fuel	Stoichiometric Flame Temp.	NOx (ppmvd/ppmvw-Methane) 1765°F/963°C – 2020°F/1104°C Firing Time	NOx (ppmvd/ppmvw-Methane) @ 15% O2, 1765°F/963°C – 2020°F/1104°C Firing Time
Methane	1.000	1.000/1.000	1.000/1.000
Propane	1.300	1.555/1.606	1.569/1.632
Butane	1.280	1.608/1.661	1.621/1.686
Hydrogen	2.067	3.966/4.029	5.237/5.299
Carbon Monoxide	2.067	3.835/3.928	4.128/0.529
Methanol	0.417-0.617	0.489/0.501	0.516/0.529
No. 2 Oil	1.667	1.567/1.647	1.524/1.614

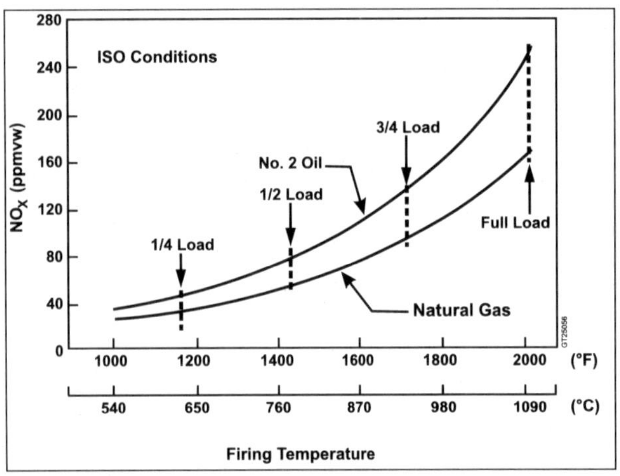

MS7001EA NO_x emissions.

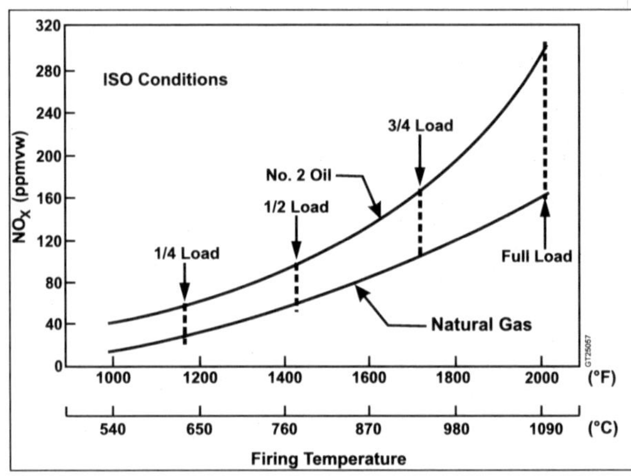

MS6001B NO_x emissions.

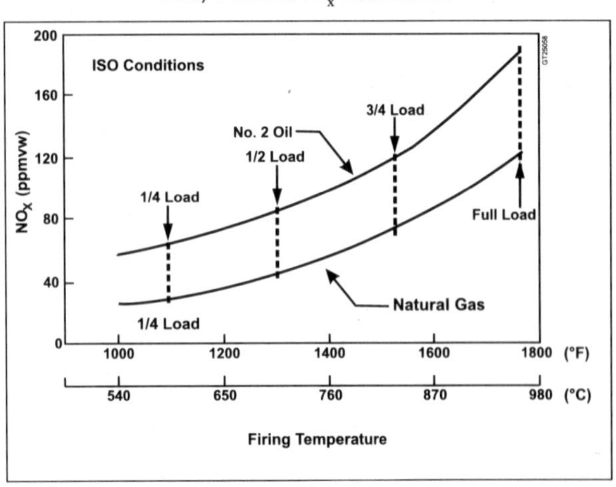

MS5001P A/T NO_x emissions.

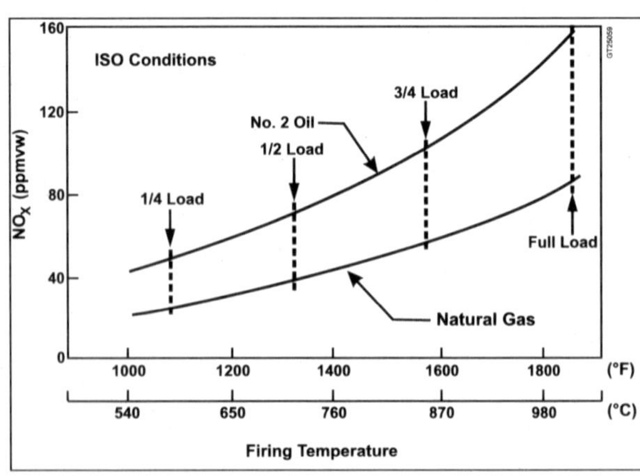

MS5001R A/T NO_x emissions.

Unburned Hydrocarbons

Unburned hydrocarbons (UHC), like carbon monoxide, are associated with combustion inefficiency. When plotted versus firing temperature, the emissions from heavy-duty gas turbine combustors show the same type of hyperbolic curve as carbon monoxide. At all but very low loads, the UHC emission levels for No. 2 distillate and natural gas are less than 7 ppmvw (parts per million by volume wet).

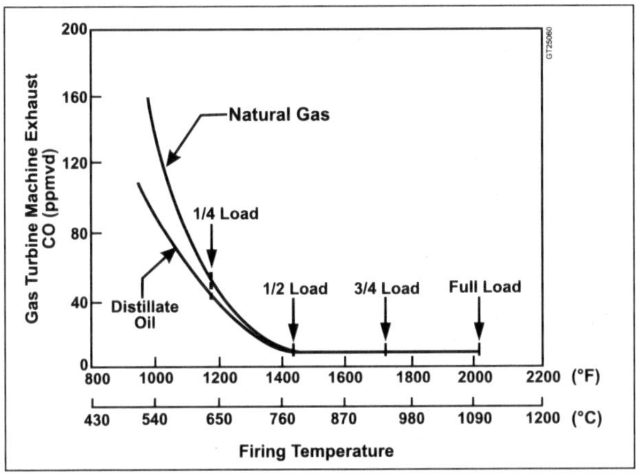

CO emissions for MS7001EA.

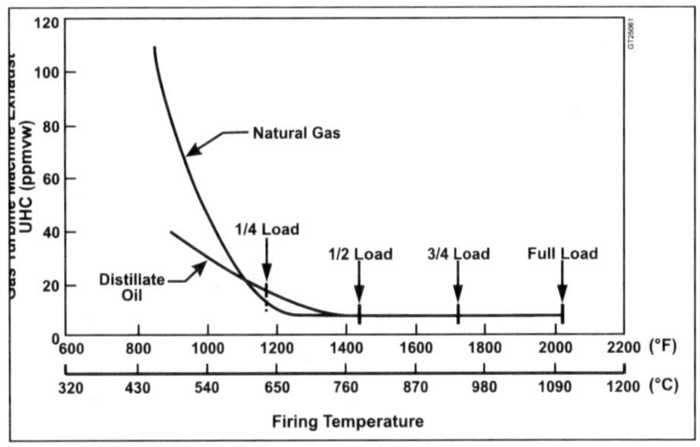

UHC emissions for MS7001EA.

Sulfur Oxides

The gas turbine itself does not generate sulfur, which leads to sulfur oxides emissions. All sulfur emissions in the gas turbine exhaust are caused by the combustion of sulfur introduced into the turbine by the fuel, air, or injected steam or water. However, since most ambient air and injected water or steam has little or no sulfur, the most common source of sulfur in the gas turbine is through the fuel. Due to the latest hot gas path coatings, the gas turbine will readily burn sulfur contained in the fuel with little or no adverse effects as long as there are no alkali metals present in the hot gas.

GE experience has shown that the sulfur in the fuel is completely converted to sulfur oxides. A nominal estimate of the sulfur oxides emissions is calculated by assuming that all fuel sulfur is converted to SO_2. However, sulfur oxide emissions are in the form of both SO_2 and SO_3. Measurements show that the ratio of SO_3 to SO_2 varies. For emissions reporting, GE reports that 95% of the sulfur into the turbine is converted to SO_2 in the exhaust. The remaining sulfur is converted into SO_3. SO_3 combines with water vapor in the exhaust to form sulfuric acid. This is of concern in most heat recovery applications where the stack exhaust temperature may be reduced to the acid dew point temperature. Additionally, it is estimated that 10% by weight of the SO_x generated is sulfur mist. By using the relationships above, the various sulfur oxide emissions can be easily calculated from the fuel flow rate and the fuel sulfur content as shown in figure.

There is currently no internal gas turbine technique available to prevent or control the sulfur oxides emissions from the gas turbine. Control of sulfur oxides emissions has typically required limiting the sulfur content of the fuel, either by lower sulfur fuel selection or fuel blending with low sulfur fuel.

Particulates

Gas turbine exhaust particulate emission rates are influenced by the design of the combustion system, fuel properties and combustor operating conditions. The principal components of the particulates are smoke, ash, ambient non-combustibles, and erosion and corrosion products. Two additional components that could be considered particulate matter in some localities are sulfuric acid and unburned hydrocarbons that are liquid at standard conditions.

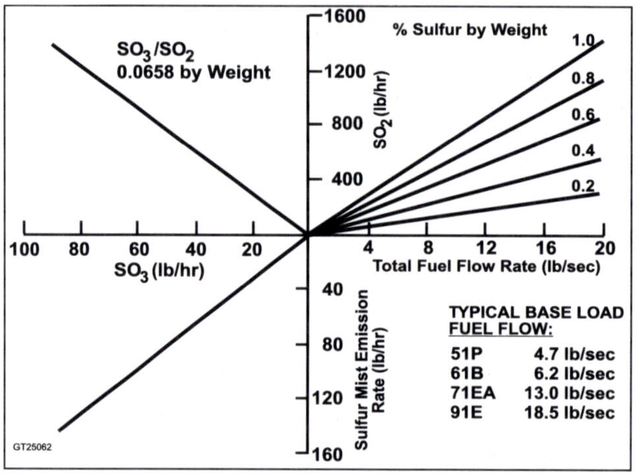

Calculated sulfur oxide and sulfur emissions.

Smoke

Smoke is the visible portion of filterable particulate material. The GE combustor design coupled with air atomization of liquid fuels has resulted in a nonvisible plume over the gas turbine load range for a wide variety of fuels. The GE smoke-measuring unit is the Von Brand Reflective Smoke Number (GEVBRSN). If this number is greater than 93 to 95 for the MS7001E, then the plume will not be visible. For liquid fuels, the GEVBRSN is a function of the hydrogen content of the fuel. For natural gas fuel, the smoke number is essentially 99 to 100 over the load range and visible smoke is not present.

Dry Emissions Estimates at Base Load

The ISO non-abated full load emissions estimates for the various GE heavy-duty gas turbine models are provided in table. The natural gas and distillate fuel emission estimates shown are for thermal NO_x, CO, UHC, VOC, and particulates. For reporting purposes, all particulates are also reported as PM-10. Therefore PM10 is not shown in the tables. The nominal full rated firing temperature for each gas turbine model is also shown in table.

As can be easily seen in the table, at base load without NO_x abatement, the emissions of CO, UHC, VOC, and particulates are quite low. The estimated values of NO_x vary between gas turbine designs and generally increase with the frame size firing temperature.

Dry Emissions Estimates at Part Load

Simple-Cycle Turbines

At turbine outputs below base load the emissions change from the values given in table. These changes are affected by the turbine configuration and application and in some cases by the turbine controls.

Single-shaft gas turbines with non-modulating inlet guide vanes operating at constant shaft speed have part load emissions characteristics which are easily estimated. For these turbines the NO_x

emissions vary exponentially with firing temperature as shown previously in figures. The load points for each turbine are also marked on these figures. Due to the conversions used in the various NO_x reporting methods, the information in figures has been redrawn in figures. This information shows the estimated ISO NO_x emissions on a ppmvd @ 15% O_2, ppmvw, and lb/hour basis for MS7001EA, MS6001B, MS5001P and MS5001R. In these figures, the nominal peak load firing temperature point is also given. It should be noted that in some cases the NO_x ppmvd@15% O_2 reporting method can cause number values to increase as load is reduced (e.g., see the MS5001P A/T in figure.) Since the GE MS9001E gas turbine is a scaled version of the MS7001E gas turbine, the MS7001E gas turbine figures can be used as an estimate of MS9001E gas turbine part load emissions characteristics. Many gas turbines have variable inlet guide vanes that are modulated closed at part load conditions in order to maintain higher exhaust temperatures for waste heat recovery equipment located in the gas turbine exhaust. As shown in figure, closing the inlet guide vanes has a slight effect on the gas turbine NO_x emissions. Figure shows the effect on NO_x ppmvd @ 15% O_2 and figure shows the effect on NO_x lb/hr. The figures show both MS5001P and MS7001E characteristics. They also show normalized NO_x (% of base load value) vs. % base load. Curves are shown for load reductions by either closing the inlet guide vanes while maintaining exhaust temperature control and for load reductions by reducing firing temperature while keeping the inlet guide vanes fully open.

Table: NOx emission levels @ 15% O2 (ppmvd).

		Dry (Non-Abated)		H_2O/Steam Inj.	
Single Shaft Units Model	Firing Temp. F/C	Gas	Dist.	Gas(FG1A/ FG1B)	Gas(FG1C/ FG1F)
MS5001P	1730/943	128	195	25	42
MS5001P-N/T	1765/963	142	211	25	42
MS6001B	2020/1104	161	279	25	65/42
MS7001B	1840/1004	109	165	25	42
MS7001B Option 3	1965/1074	124	191	25	42
MS7001B Option 4	2020/1104	132	205	25	42
MS7001EA	2020/1104	160	245	25	42
MS9001B	1940/1060	109	165	42	65
MS9001B Option 3	1965/1074	124	191	42	65
MS9001B Option 4	2020/1104	132	205	42	65
MS9001E	2020/1104	157	235	42	65
MS9001E	2055/1124	162	241	42	65
6FA	2350/1288				
7FA	2400/1316				
7FA	2420/1327				
9FA	2350/1288				
		Dry (Non-Abated)		H_2O/Steam Inj.	
Two Shaft Units* Model	Firing Temp. F/C	S.C.	R.C.**	S.C.	S.C.
MS3002F	1575/1625/857/885	115	201	42	50
MS3002J	1730/943	128	217	42	50

MS3002J-N/T	1770/968	140	236	42	50
MS5002	1700/927	125	220	42	50
MS5002B-N/T	1770/966	137	255	42	50

Mechanical drive gas turbines typically vary the output load shaft speed in order to adjust the turbine output to match the load equipment characteristic. Single-shaft gas turbines operating on exhaust temperature control have a maximum output NO_x emissions characteristic vs. turbine shaft speed, as shown in figure for an MS5001R Advanced Technology uprated turbine. The characteristic shown is primarily due to the gas turbine exhaust temperature control system and the turbine thermodynamics. As seen in figure, as the turbine output shaft speed is reduced below 100%, NO_x emissions decrease directly with turbine shaft speed. As the speed decreases, the exhaust temperature increases till the exhaust component temperature limit is reached. Once the exhaust isothermal limit is reached, the variation of NO_x emissions with speed will become greater. In figure this exhaust isothermal temperature limit is reached at approximately 84% speed. Two-shaft gas turbines also vary the output turbine shaft speed with load conditions. However the gas turbine compressor shaft and combustor operating conditions are controlled independent of the output shaft speed. On a two-shaft gas turbine, if the gas turbine compressor shaft speed is held constant by the control system while on exhaust temperature control, the NO_x emissions are not affected by the load turbine shaft speed.

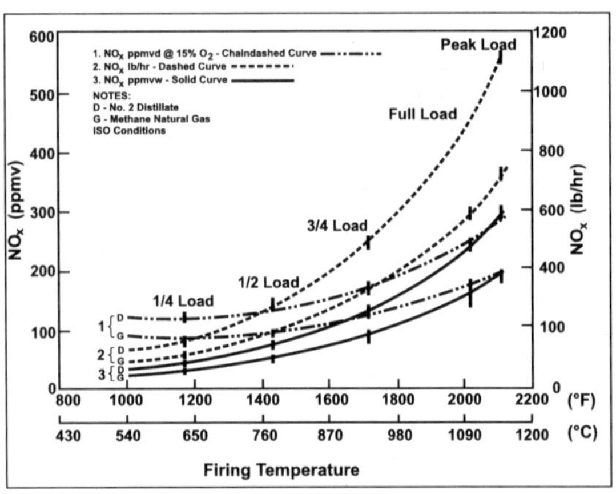

MS7001EA NO_x emissions.

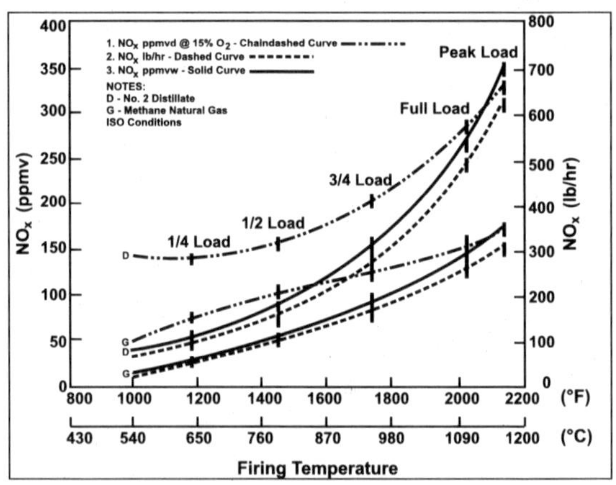

MS6001B NO_x emissions.

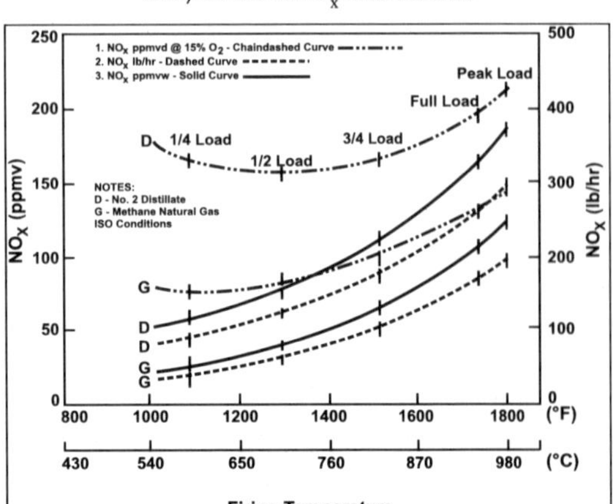

MS5001P A/T NO_x emissions.

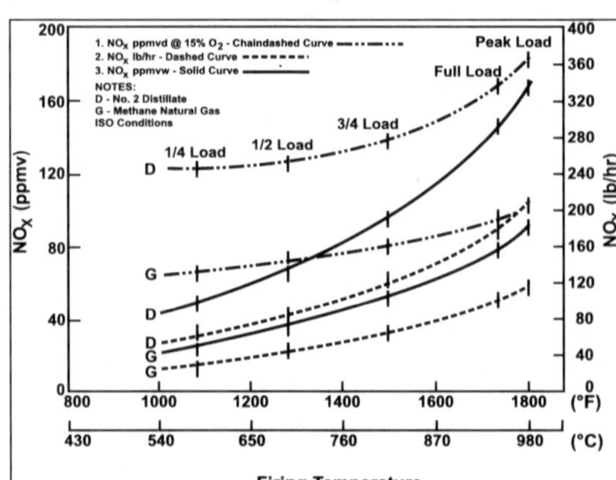

MS5001R A/T NO_x emissions.

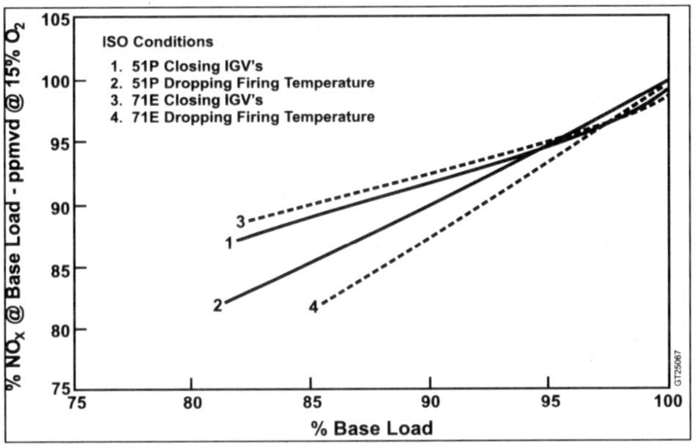

Inlet guide vane effect on NO_x ppmvd @ 15% O_2 vs. load.

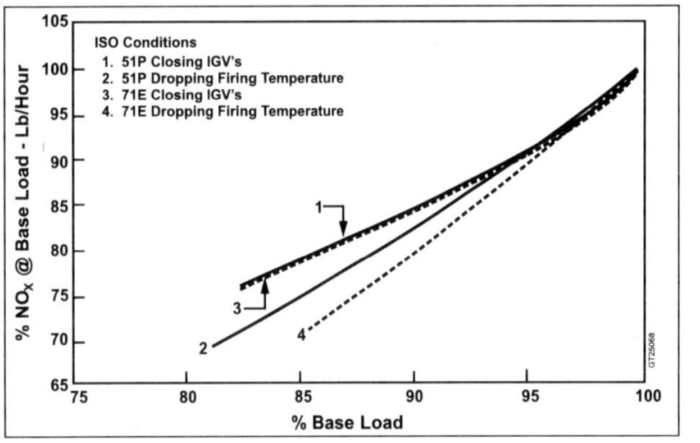

Inlet guide vane effect on NO_x lb/hour vs. load.

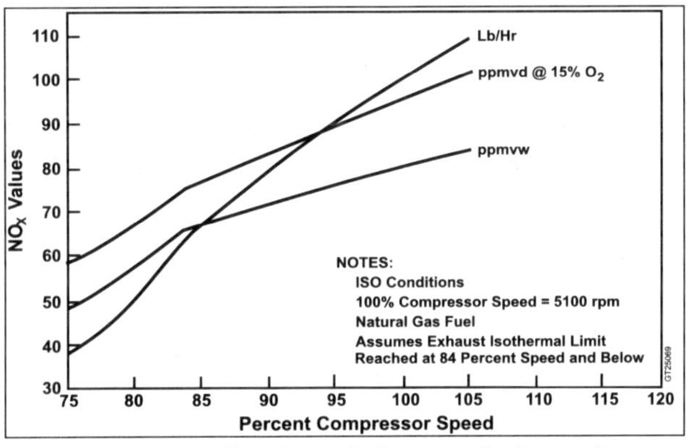

MS5001R A/T NO_x emissions vs. shaft speed.

Exhaust Heat Recovery Turbines

Regenerative cycle and waste heat recovery twoshaft gas turbines are normally controlled to operate the gas turbine compressor at the minimum speed allowable for the desired load output. As load is increased from minimum, the gas turbine compressor speed is held at minimum until the turbine exhaust temperature reaches the temperature control curve. With further increase in load, the control system will increase the gas turbine compressor speed while following the exhaust temperature control curve. If the turbine has modulated inlet guide vanes, the inlet guide vanes will open first when the exhaust temperature control curve is reached, and then, once the inlet guide vanes are fully open, the gas turbine compressor speed will be increased.

Figure shows the NO_x characteristic of a regenerative cycle MS3002J gas turbine at ISO conditions. Initially, as load is increased, NO_x increases with firing temperature while the gas turbine compressor is operating at minimum speed. For the turbine shown, the exhaust isothermal temperature control is reached at approximately 48% load. The gas turbine compressor shaft speed is then increased by the control system for further increases in load up to the 100% load point. At approximately 96% load, the gas turbine exhaust temperature control curve begins to limit exhaust temperature below the isothermal exhaust temperature due to the increasing airflow through the turbine and the NO_x values are reduced by the characteristic shown.

For a typical regenerative cycle MS5002B Advanced Technology gas turbine with modulated inlet guide vanes, the curve of NO_x vs. load at ISO conditions is shown in figure.

The NO_x vs. load characteristic is similar to the MS3002J. However, this design turbine will operate at low load with the inlet guide vanes partially closed and at minimum operating gas turbine compressor shaft speed. During initial loading, NO_x increases with firing temperature. When the exhaust temperature control system isothermal temperature limit is reached the inlet guide vanes are modulated open as load is increased. At approximately 90% load the gas turbine exhaust temperature control curve begins to limit exhaust temperature below the isothermal exhaust temperature due to the increasing airflow through the turbine and the NO_x values are reduced. At approximately 91.5% load for this turbine calculation, the inlet guide vanes are fully open and further increases in load are accomplished by increasing the gas turbine compressor speed resulting in the NO_x reduction as shown.

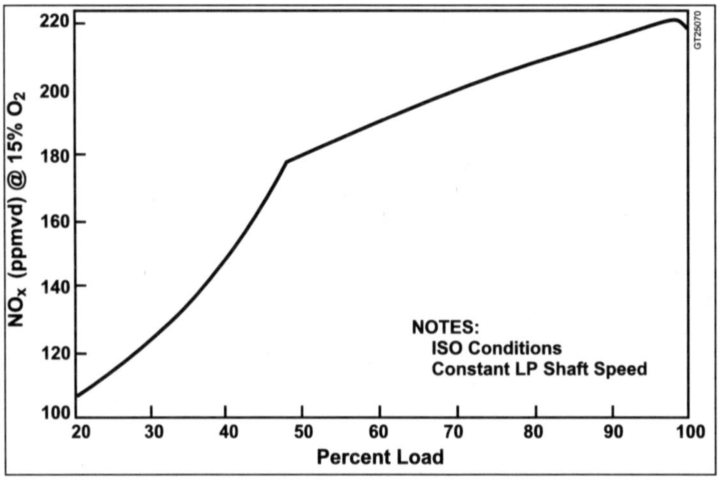

MS3002J regenerative NO_x vs. load.

Other NOx Influences

The previous sections of this paper consider the internal gas turbine design factors which influence emissions generation. There are many external factors to the gas turbine which impact the formation of NO_x emissions in the gas turbine cycle. Some of these factors will be discussed below. In all figures under this topic, the NO_x is presented as a percentage value where 100% represents the thermal ISO NO_x value for the turbine operating on base temperature control. For all figures except for the regenerator changes discussed, the curves drawn represent a single "best fit" line through the calculated characteristics for frame 3, 5, 6, 7, and 9 gas turbines. However, the characteristics shape that is shown is the same for all turbines.

Ambient Pressure: NO_x ppm emissions vary almost directly with ambient pressure. figure provides an approximation for the ambient pressure effect on NOx production on a lb/hr basis and on a ppmvd @ 15% O_2 basis. This figure is at constant 60% relative humidity. It should be noted that specific humidity varies with ambient pressure and that this variation is also included in the figure curves.

Ambient Temperature: Typical NO_x emissions variation with ambient temperature is shown in figure. This figure is drawn at constant ambient pressure and 60% relative humidity with the gas turbine operating constant gas turbine firing temperature. For an operating gas turbine the actual NO_x characteristic is directly influenced by the control system exhaust temperature control curve, which can change the slope of the curves. The typical exhaust temperature control curve used by GE is designed to hold constant turbine firing temperature in the 59 °F/15 °C to 90 °F/32 °C ambient temperature range. The firing temperature with this typical curve causes under-firing of approximately 20 °F/11 °C at 0 °F/−18 °C ambient, and approximately 10 °F/6 °C under-firing at 120 °F/49 °C ambient. Factors such as load limits, shaft output limits, and exhaust system temperature limits are also not included in the figure curves. Based on the actual turbine exhaust temperature control curve used and other potential limitations that reduce firing temperature, the estimated NO_x emissions for an operating gas turbine are typically less than the values shown in figure at both high and low ambients.

Relative Humidity: This parameter has a very strong impact on NO_x. The ambient relative humidity effect on NO_x production at constant ambient pressure of 14.7 psia and ambient temperatures of 59°F/15°C and 90°F/32°C is shown in figure.

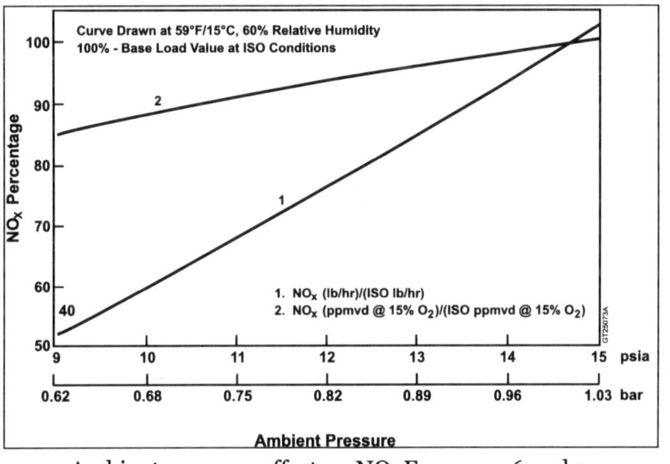

Ambient pressure effect on NO_x Frames 5, 6 and 7.

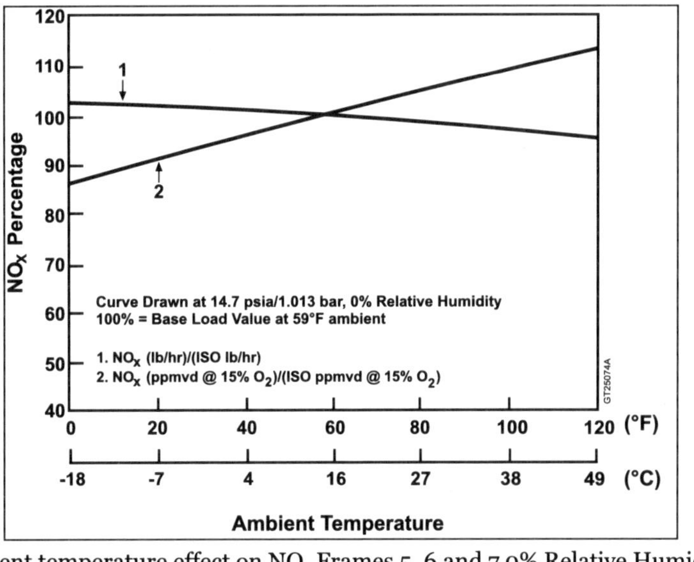

Ambient temperature effect on NO_x Frames 5, 6 and 7 0% Relative Humidity.

The impact of other parameters such as inlet/exhaust pressure drops, regenerator characteristics, evaporative/inlet coolers, etc., are similar to the ambient parameter effects described above. Since these parameters are usually unit specific, customers should contact GE for further information.

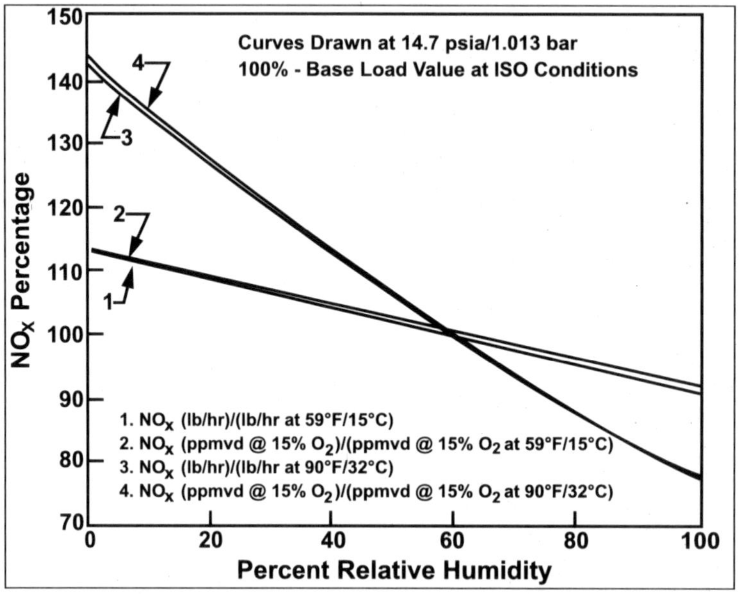

Relative humidity effect on NO_x Frames 5, 6 and 7.

Power Augmentation Steam Injection: The effect of power augmentation steam injection on gas turbine NO_x emissions is similar to NO_x steam injection on a ppmvw and lb/hr basis. However, only approximately 30% of the power augmentation steam injected participates in NO_x reduction. The remaining steam flows through dilution holes downstream of the NO_x producing area of the combustor. 100% of the power augmentation steam injected is used in the conversion from ppmvw to ppmvd @ 15% O_2.

Emission Reduction Techniques

The gas turbine, generally, is a low emitter of exhaust pollutants because the fuel is burned with ample excess air to ensure complete combustion at all but the minimum load conditions or during start-up. The exhaust emissions of concern and the emission control techniques can be divided into several categories as shown in table. Each pollutant emission reduction technique will be discussed in the following sections.

Nitrogen Oxides Abatement

The mechanism on thermal NO_x production was first postulated by Zeldovich. This is shown in figure. It shows the flame temperature of distillate as a function of equivalence ratio. This ratio is a measure of fuel-to-air ratio in the combustor normalized by stoichiometric fuel-to-air ratio. At the equivalence ratio of unity, the stoichiometric conditions are reached. The flame temperature is highest at this point. At equivalence ratios less than 1, we have a "lean" combustor. At the values greater than 1, the combustor is "rich." All gas turbine combustors are designed to operate in the lean region.

Figure shows that thermal NO_x production rises very rapidly as the stoichiometric flame temperature is reached. Away from this point, thermal NO_x production decreases rapidly. This theory then provides the mechanism of thermal NO_x control. In a diffusion flame combustor, the primary way to control thermal NO_x is to reduce the flame temperature.

Table: Emission control techniques.

NO_x	Lean Head End Liner Water or Steam Injection Dry Low NO_x
CO	Combustor Design Catalytic Reduction
UHC & VOC	Combustor Design
SO_x	Control Sulfur in Fuel
Particulates & PM-10	Fuel Composition
Smoke Reduction	Combustor Design • Fuel Composition • Air Atomization
Particulate Reduction	Fuel Composition • Sulfur • Ash

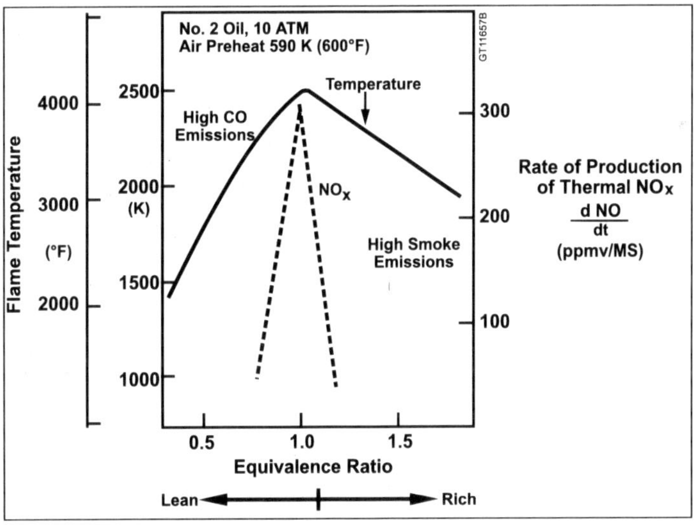

NO_x production rate.

Lean Head End (LHE) Combustion Liners

Since the overall combustion system equivalence ratio must be lean (to limit turbine inlet temperature and maximize efficiency), the first efforts to lower NO_x emissions were naturally directed toward designing a combustor with a leaner reaction zone. Since most gas turbines operate with a large amount of excess air, some of this air can be diverted towards the flame end, which reduces the flame temperature.

Leaning out the flame zone (reducing the flame zone equivalence ratio) also reduces the flame length, and thus reduces the residence time a gas molecule spends at NO_x formation temperatures. Both these mechanisms reduce NO_x. The principle of a LHE liner design is shown in figure.

It quickly became apparent that the reduction in primary zone equivalence ratio at full operating conditions was limited because of the large turndown in fuel flow (40 to 1), air flow (30 to 1), and fuel/air ratio (5 to 1) in industrial gas turbines. Further, the flame in a gas turbine is a diffusion flame since the fuel and air are injected directly into the reaction zone. Combustion occurs at or near stoichiometric conditions, and there is substantial recirculation within the reaction zone. These parameters essentially limit the extent of LHE liner technology to a NO_x reduction of 40% at most. Depending upon the liner design, actual reduction achieved varies from 15% to 40%.

Figure compares an MS5001P LHE liner to a standard liner. The liner to the right is the LHE liner. It has extra holes near the head (flame) end and also has a different louver pattern compared to the standard liner. Table summarizes all LHE liners designed to date. Field test data on MS5002 simple-cycle LHE liners and MS3002J simple-cycle LHE liners are shown in figures.

One disadvantage of leaning out the head end of the liner is that the CO emissions increase. This is clear from figure, which compares CO between the standard and LHE liner for a MS5002 machine.

Water/Steam Injection

Another approach to reducing NO_x formation is to reduce the flame temperature by introducing a heat sink into the flame zone. Both water and steam are very effective at achieving this goal. A penalty in overall efficiency must be paid for the additional fuel required to heat the water to combustor temperature. However, gas turbine output is enhanced because of the additional mass flow through the turbine. By necessity, the water must be of boiler feed water quality to prevent deposits and corrosion in the hot turbine gas path area downstream of the combustor.

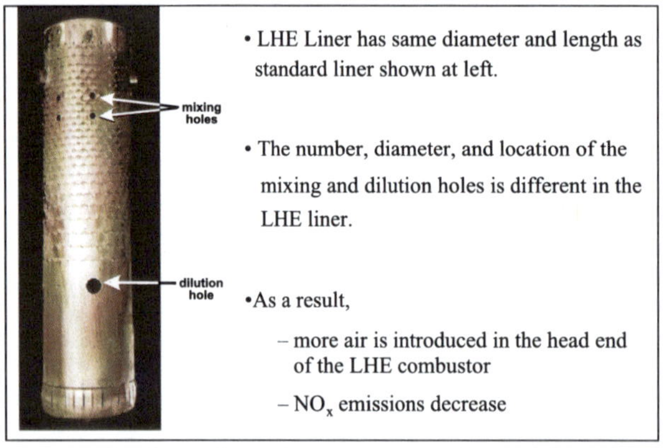

Standard simple-cycle MS5002 combustion liner.

Louvered low NO_x lean head end combustion liners.

Water injection is an extremely effective means for reducing NO_x formation; however, the combustor designer must observe certain cautions when using this reduction technique. To maximize the effectiveness of the water used, fuel nozzles have been designed with additional passages to inject water into the combustor head end. The water is thus effectively mixed with the incoming combustion air and reaches the flame zone at its hottest point. In figure 26 the NO_x reduction achieved by water injection is plotted as a function of water-to-fuel ratio for an MS7001E machine. Other machines have similar NO_x abatement performance with water injection.

Steam injection for NO_x reduction follows essentially the same path into the combustor head end as water. However, steam is not as effective as water in reducing thermal NO_x. The high latent heat of water acts as a strong thermal sink in reducing the flame temperature.

In general, for a given NO_x reduction, approximately 1.6 times as much steam as water on a mass basis is required for control.

There are practical limits to the amount of water or steam that can be injected into the combustor before serious problems occur. This has been experimentally determined and must be taken into account in all applications if the combustor designer is to ensure long hardware life for the gas turbine user.

Table: Lean head end (LHE) liner development.

Turbine Model	Laboratory Development Completed	First Field Test
S/C MS3002F	December-98	Fall 1999
S/C MS3002G	December-98	to be determined
S/C MS3002J	April-97	March-99
S/C MS5002B, C, & D	April-97	September-97
S/C MS5001 (All Models)	1986	Over 130 operating in field
R/C MS3002J	February-99	to be determined
R/C MS5002B & C	February-99	to be determined

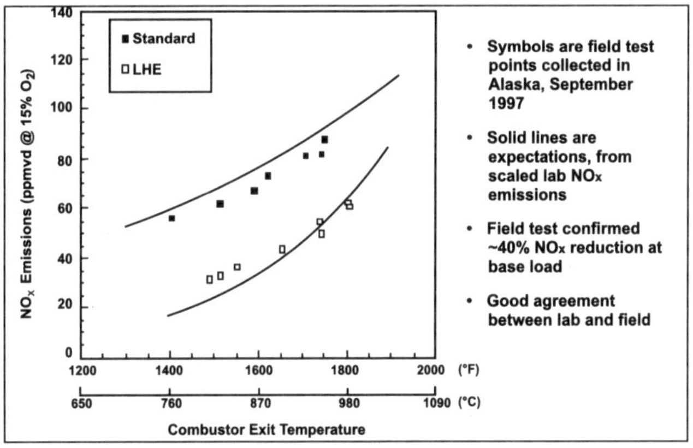

Field test data: simple-cycle MS5002 NO_x.

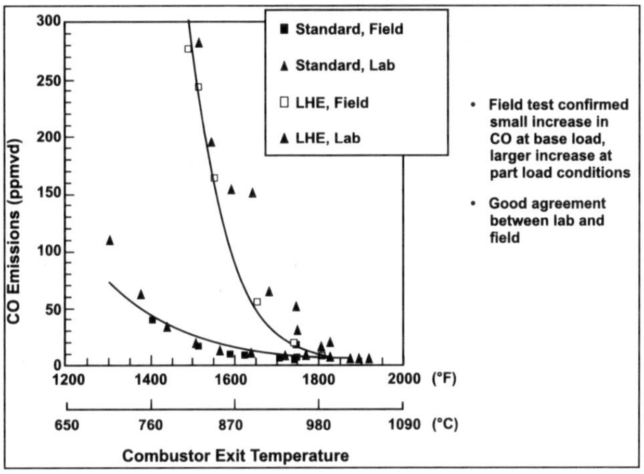

Field test data: simple-cycle MS5002 CO.

Injecting water/steam in a combustor affects several parameters:

- Dynamic Pressure Activity within the Combustor: Dynamic pressures can be defined as pressure oscillations within the combustor driven by non-uniform heat release rate inherent in any diffusion flame or by the weak coupling between heat release rate, turbulence, and acoustic modes. An example of the latter is selective amplification of combustion roar by the acoustic modes of the duct. Frequencies range from near zero to several hundred hertz. Figure shows dynamic pressure activity for both water injection and steam injection for an MS7001E combustor. Water injection tends to excite the dynamic activity more than steam injection. The oscillating pressure loads on the combustion hardware act as vibratory forcing functions and therefore must be minimized to ensure long hardware life. Through combustor design modifications such as the addition of a multi-nozzle fuel system, significant reductions in dynamic pressure activity are possible.

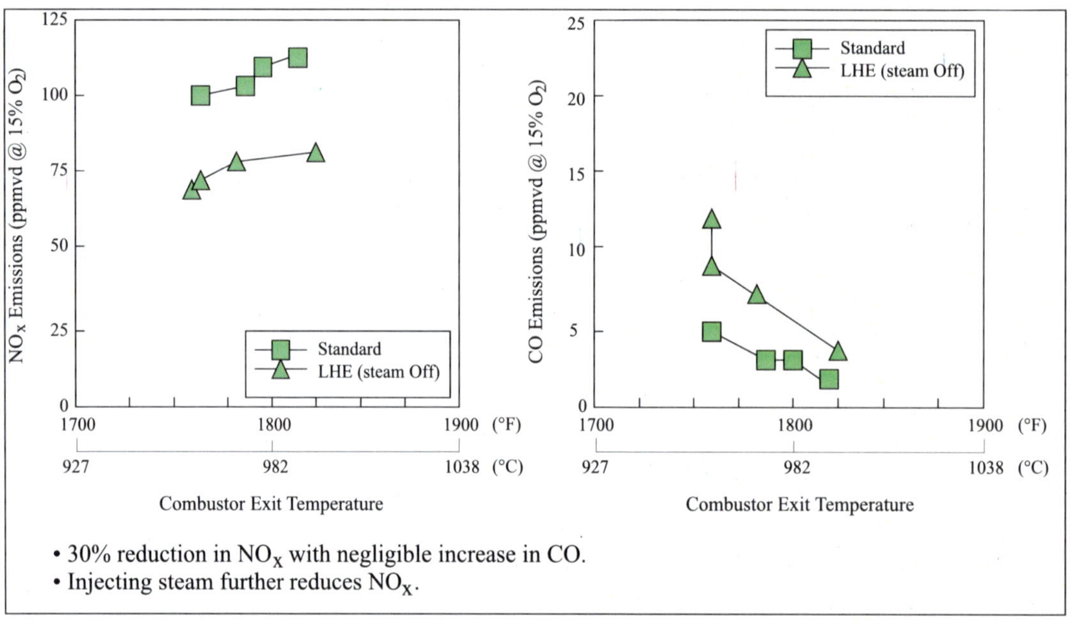

Field test data: simple-cycle MS3002J with steam injection for power augmentation.

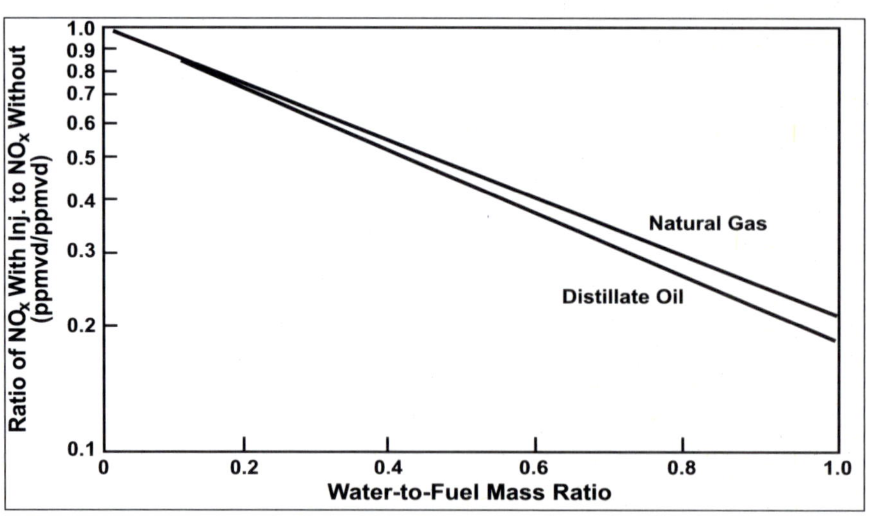

MS7001E NO_x reduction with water injection.

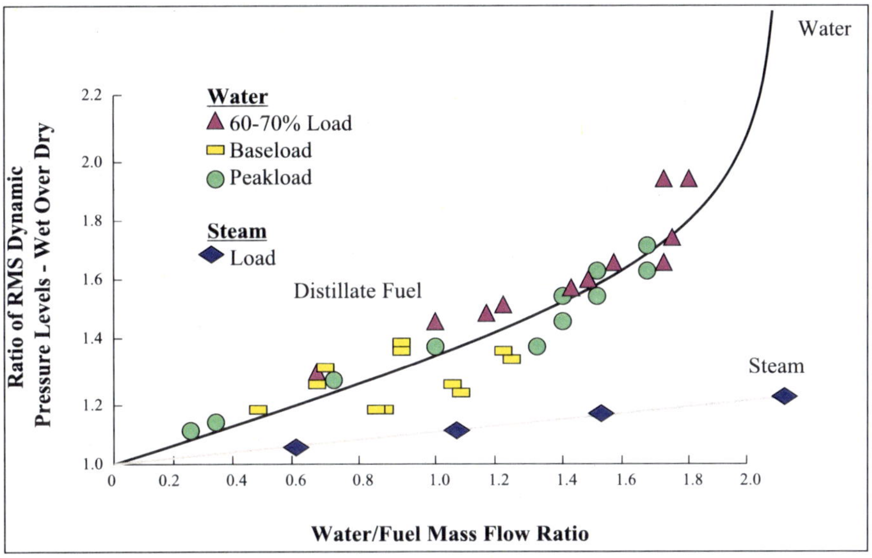

MS7001E combustor dynamic pressure activity.

- Carbon Monoxide Emissions: As more and more water/steam is added to the combustor, a point is reached at which a sharp increase in carbon monoxide is observed. This point has been dubbed the "knee of the curve". Once the knee has been reached for any given turbine inlet temperature, one can expect to see a rapid increase in carbon monoxide emissions with the further addition of water or steam. Obviously, the higher the turbine inlet temperature, the more tolerant the combustor is to the addition of water for NO_x control. Figure shows the relationship of carbon monoxide emissions to water injection for a MS7001B machine for natural gas fuel. Figure shows the effect of steam injection on CO emissions for a typical MS7001EA. Unburned hydrocarbons have a similar characteristic with NO_x water or steam injection as carbon monoxide. Figure shows the MS7001EA gas turbine unburned hydrocarbon versus firing temperature characteristic with steam injection.

- Combustion Stability: Increasing water/steam injection reduces combustor-operating stability.

- Blow Out: With increasing water/steam injection, eventually a point will be reached when the flame will blow out. This point is the absolute limit of NO_x control with water/steam injection.

Carbon Monoxide Control

There are no direct carbon monoxide emission reduction control techniques available within the gas turbine. Basically the carbon monoxide emissions within the gas turbine combustor can be viewed as resulting from incomplete combustion. Since the combustor design maximizes combustion efficiency, carbon monoxide emissions are minimized across the gas turbine load range of firing temperatures. Reviewing figure shows that the carbon monoxide emission levels increase at lower firing temperatures. In some applications where carbon monoxide emissions become a concern at low loads (firing temperatures), the increase in carbon monoxide can be lowered by.

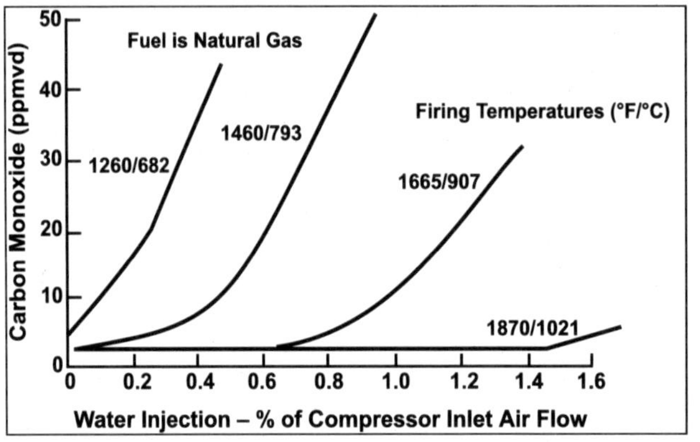

Carbon monoxide vs. water injection effect of firing temperature – MS7001B.

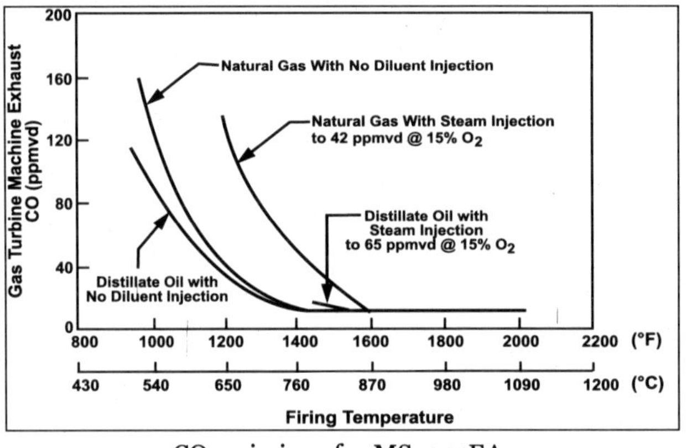

CO emissions for MS7001EA.

- Reducing the amount of water/steam injection for NO_x control (if allowed).
- Closing the inlet guide vanes, which will increase the firing temperature for the same load.

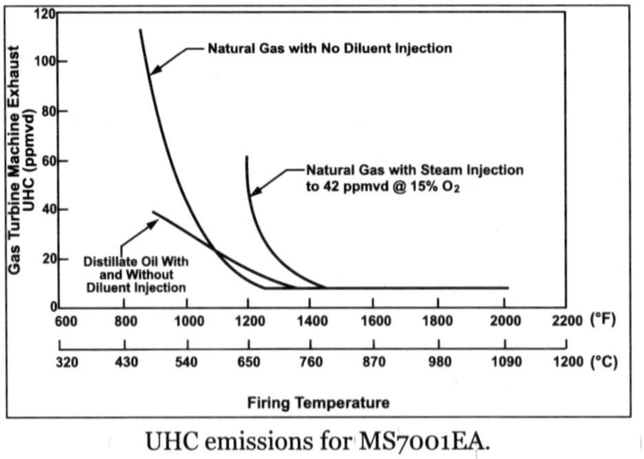

UHC emissions for MS7001EA.

Unburned Hydrocarbons Control

Similar to carbon monoxide, there are also no direct UHC reduction control techniques used within the gas turbine. UHCs are also viewed as incomplete combustion, and the combustor is designed to minimize these emissions. The same indirect emissions control techniques can be used for unburned hydrocarbons as for carbon monoxide.

Particulate and Smoke Reduction

Control techniques for particulate emissions with the exception of smoke are limited to control of the fuel composition.

Although smoke can be influenced by fuel composition, combustors can be designed which minimize emission of this pollutant. Heavy fuels such as crude oil and residual oil have low hydrogen levels and high carbon residue, which increase smoking tendencies. GE has designed heavy-fuel combustors that have smoke performance comparable with those which burn distillate fuel.

Crude and residual fuel oil generally contain alkali metals (Na, K) in addition to vanadium and lead, which cause hot corrosion of the turbine nozzles and buckets at the elevated firing temperatures of today's gas turbine. If the fuel is washed, water soluble compounds (alkali salts) containing the contaminants are removed. Filtration, centrifuging, or electrostatic precipitation are also effective on reducing the solid contaminants in the combustion products.

Contaminants that cannot be removed from the fuel (vanadium compounds) can be controlled through the use of inhibitors. GE uses addition of magnesium to control vanadium corrosion in its heavy-duty gas turbines. These magnesium additives always form ash within the hot gas path components. This process generally requires control and removal of added ash deposits from the turbine. The additional ash will contribute to the exhaust particulate emissions. Generally, the expected increase can be calculated from an analysis of the particular fuel being burned.

In some localities, condensable compounds such as SO_3 and condensable hydrocarbons are considered particulates. SO_3, like SO_2, can best be minimized by controlling the amount of sulfur in the fuel. The major problem associated with sulfur compounds in the exhaust comes from the difficulty of measurement. Emissions of UHCs, which are a liquid or solid at room temperature, are very low and only make a minor contribution to the exhaust particulate loading.

Water/Steam Injection Hardware

The injection of water or steam into the combustion cover/fuel nozzle area has been the primary method of NO_x reduction and control in GE heavy-duty gas turbines since the early 1970s. The same design gas turbine equipment is supplied for conversion retrofits to existing gas turbines for either injection method. Both NO_x control injection methods require a microprocessor controller, therefore turbines with older controls need to have their control system upgraded to Mark V or Mark VI SPEEDTRONIC™ controls conversion. The control system for both NO_x control injection methods utilizes the standard GE gas turbine control philosophy of two separate independent methods for shutting off the injection flow.

The NO_x water injection system is shown schematically in figure and consists of a water pump and filter, water flow meters, water stop and flow control valves. This material is supplied on a skid approximately 10 x 20 feet in size for mounting at the turbine site. The water from the skid is piped to the turbine base where it is manifold to each of the fuel nozzles using pigtails. The water injection at the combustion chamber is through passages in the fuel nozzle assembly. A typical water injection fuel nozzle assembly is shown schematically in figure. For this nozzle design there are eight or twelve water spray nozzles directing the water injection spray towards the fuel nozzle tip swirler. While this design is quite effective in controlling the NO_x emissions, the water spray has a tendency to impinge on the nozzle tip swirler and on the liner cap/cowl assembly. Resulting thermal strain usually leads to cracks, which limits the combustion inspections to 8000 hours or less. To eliminate this cracking, the latest design water-injected fuel nozzle is the breech-load fuel nozzle. In this design the water is injected through a central fuel nozzle passage, injecting the water flow directly into the combustor flame. Since the water injection spray does not impinge on the fuel nozzle swirler or the combustion cowl assembly, the breech load fuel nozzle design results in lower maintenance and longer combustion inspection intervals for NO_x water injection applications.

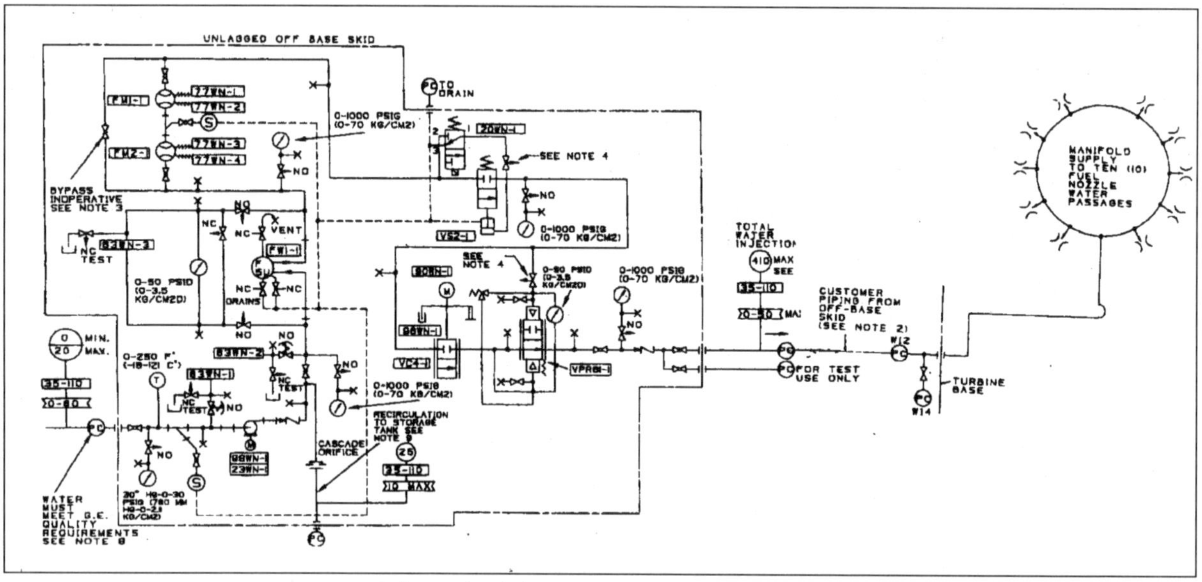

Schematic piping – water injection system.

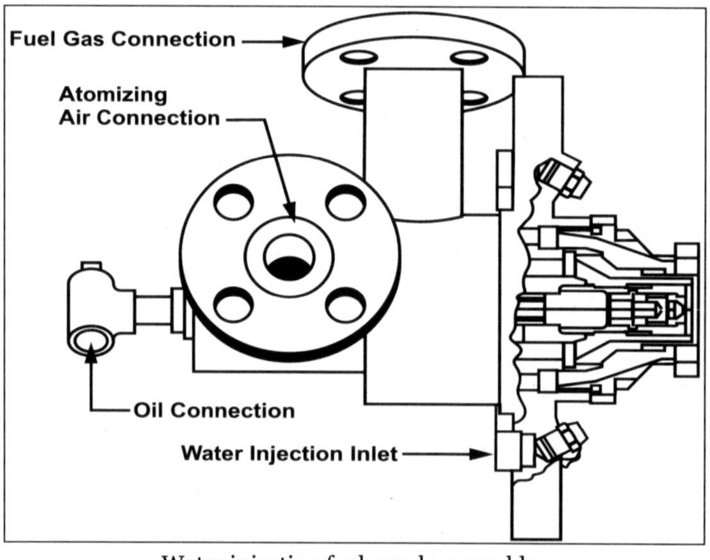

Water injection fuel nozzle assembly.

The NO_x steam injection system is shown schematically in figure, and consists of a steam flow meter, steam control valve, steam stop valve, and steam blow down valves. This material is supplied loose for mounting near the turbine base by the customer. The steam injection flow goes to the steam-injection manifold on the turbine base. Flexible pigtails are used to connect from the steam manifold to each combustion chamber. The steam injection into the combustion chamber is through machined passages in the combustion can cover. A typical steam-injection combustion cover with the machined steam-injection passage and steam injection nozzles is shown in figure.

Water quality is of concern when injecting water or steam into the gas turbine due to potential problems with hot gas path corrosion, and effects to the injection control equipment. The injected water or steam must be clean and free of impurities and solids. The general requirements of the injected water or steam quality are shown in table. Total impurities into the gas turbine are a total of the ambient air, fuel, and injected water or steam. The total impurities requirement may lower the water or steam injection quality requirements. It is important to note that the total impurities requirement is provided relative to the input fuel flow.

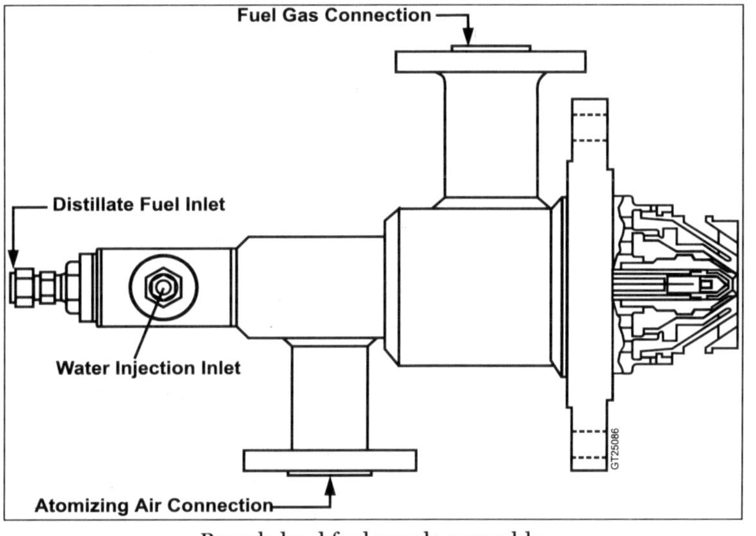

Breech-load fuel nozzle assembly.

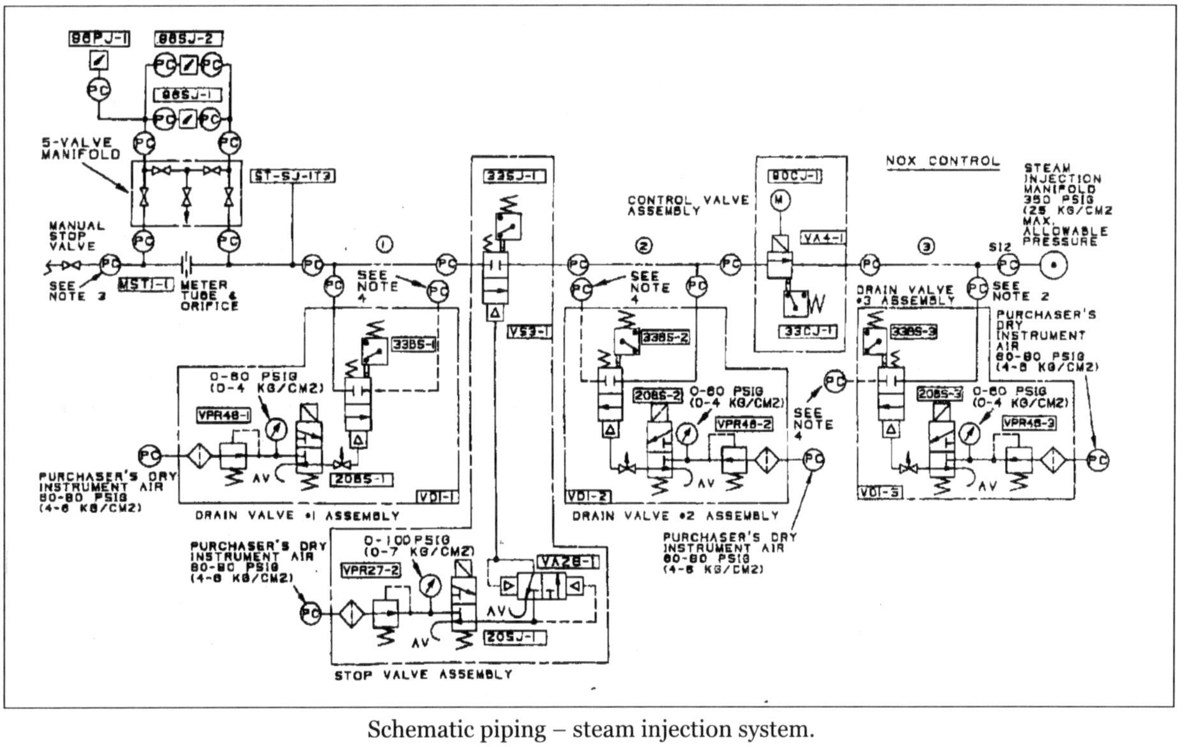

Schematic piping – steam injection system.

Minimum NO_x Levels

As described above, the methods used to reduce thermal NO_x inside the gas turbine are by combustor design or by diluent injection. NO_x emissions from each frame size without any control, refer to table. With the LHE liner design, dry (no water/steam injection) NO_x emissions could be reduced by 15–40% relative to standard liner. This is the limit of LHE liner technology.

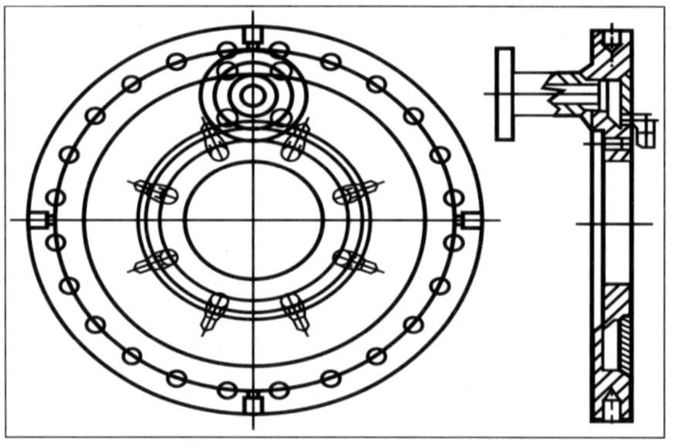

Combustion cover – steam injection.

Table: Water or steam injection quality requirements.

Water/Steam Quality Total Dissolved Solids (Sodium + Potassium + Vanadium + Lead) pH	5.0 ppm Max. 0.5 ppm Max. 6.5 – 7.5
Total Limits in all Sources (Fuel, Steam, Water, Air) Contaminant Sodium + Potassium Lead Vanadium Calcium	Max. Equivalent Concentration (ppm – wt) 1.0 1.0 0.5 2.0

With water or steam injection, significant reduction in NO_x is achieved. The lowest achievable NO_x values with water/steam injection from GE heavy-duty gas turbines are also shown in table. The table provides the current minimum NO_x levels for both methane natural gas fuel and #2 distillate fuel oil.

Maintenance Effects

As described previously, the methods used to control gas turbine exhaust emissions have an effect on the gas turbine maintenance intervals. Table provides the recommended combustion inspection intervals for current design Advanced Technology combustion systems used in base load continuous duty gas turbines without NO_x control systems and the recommended combustion inspection intervals with the various NO_x control methods at the NO_x ppmvd @ 15% O_2 levels shown. Both natural gas fuel and distillate fuel recommended combustion inspection intervals are included. Review of table shows that the increased combustion dynamics (as the combustor design goes from dry to steam injection) and then to water injection results in reductions in the recommended combustion inspection intervals.

Performance Effects

As mentioned previously the control of NO_x can impact turbine firing temperature and result in gas turbine output changes. Additionally, the injection of water or steam also impacts gas turbine output, heat rate, and exhaust temperature. Figure shows the impact of NO_x injection on these gas turbine parameters when operating at base load for all single shaft design gas turbines. Note that the injection rate is shown as a percentage of the gas turbine compressor inlet airflow on a weight basis. The output and heat rate change is shown on a percent basis while exhaust temperature is shown in degrees F. Review of figure shows that turbine output is increased when NO_x injection is used. The gas turbine load equipment must also be capable of this output increase or control changes must be made in order to reduce the gas turbine output.

	Natural gas/ No.2 Distillate ppmvd @15% O2	Natural Gas Fired Hours of Operation Water/Steam Injection	No. 2 Distillate Fired Hours of Operation Water/Steam Injection
MS5001P N/T Dry	142/211	12,000/12,000	12,000/12,000
NSPS	87/86	12,000/12,000	6,000/6,000
	42/65	6,000/6,000	6,000/6,000
	42/42	6,000/6,000	1,500/4,000
MS6001B Dry	148/267	12,000/12,000	12,000/12,000
NSPS	94/95	8,000/8,000	6,000/6,000
	42/65	8,000/8,000	8,000/8,000
	42/42	8,000/8,000	4,000/4,000
MS7001E Dry	154/228	8,000/8,000	8,000/8,000
NSPS	96/97	8,000/8,000	8,000/8,000
	42/65	6,500/8,000	6,500/8,000
	42/42	6,500/8,000	1,500/3,000
	25/42	8,000/8,000	6,000/6,000
MS9001E Dry	147/220	8,000/8,000	8,000/8,000
	42/65	6,500/8,000	6,500/8,000
Inspection Intervals reflect current hardware. Older units with earlier vintage hardware will have lower Inspection intervals. The above values represent initial recommended combustion inspection intervals. The intervals are subject to change based on experience. Base Load Operation. NSPS NO_x levels are 75 ppm with heat rate correction included.			

Table: Estimated ISO NO_x level effects on combustion inspection intervals.

The emissions characteristics of gas turbines have been presented both at base load and part load conditions. The interaction of emission control on other exhaust emissions as well as the effects on gas turbine maintenance and performance have also been presented. The minimum controllable NO_x levels using LHE and water/steam injection techniques have also been presented. Using this information, emissions estimates and the overall effect of the various emission control methods can be estimated.

GAS TURBINE NOISE

In a gas turbine, there are three main sources of noise such as intake, exhaust, and casing. Intake noise is created by the interaction of the axial air compressor rotor and stator, and is a function of blade number, tip speed, and pressure increase. Intake noise is less than exhaust noise in overall, but its high frequency content sounds are much larger than that of the exhaust noise. Exhaust noise has higher amplitude and has lower frequency due to combustion process. Typically, the inlet and exhaust sound power levels range from 120 dB to over 155 dB. Casing noise is generated through high speed misaligned mechanical components in the turbine housing radiating to the outer casing. In principle, gas turbine noise come from aerodynamic sources. High aerodynamic turbulence are combustion are present in the operation of gas turbine. Combustion would be a monopole source of sound along with rotating shock waves. Dipole sources of sound would mainly be from fluctuating forces on blades and guide vanes, and free jets would be a quadrupole noise.

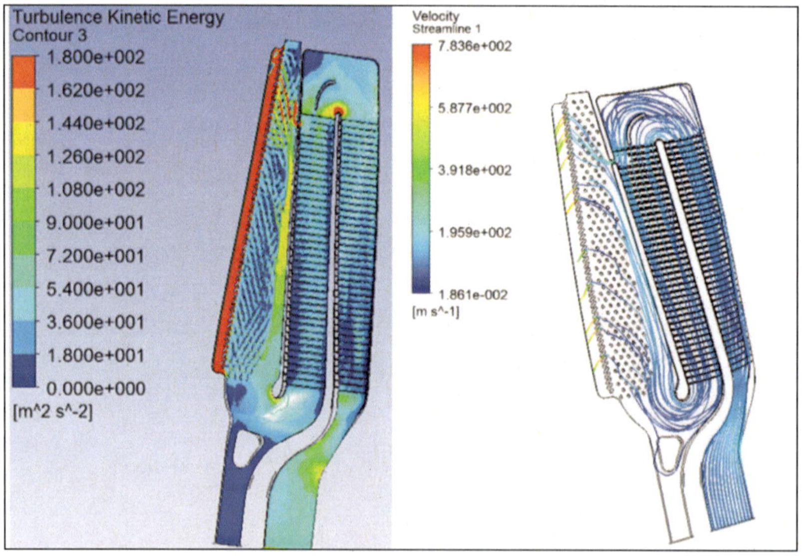

COMBUSTION NOISE

Combustion noise is generated inside the combustion chamber and has to propagate through turbine blades. The problem of acoustic wave propagation through accelerating flows was first studied by Tsien in a rocket engine combustion instability framework. It was Marble and Candel, however, who first included the propagation of entropy waves through accelerating subsonic and supersonic nozzles, explaining the two combustion noise generation mechanisms: direct and indirect, as seen in figure. Direct noise is generated by the acoustic waves produced in the flame region, which travel downstream through the turbine stages to the outlet. Indirect combustion noise is generated by the entropy waves induced by the unsteady combustion process. These waves generate noise when propagating through non-homogeneous flows, such as the turbine stages.

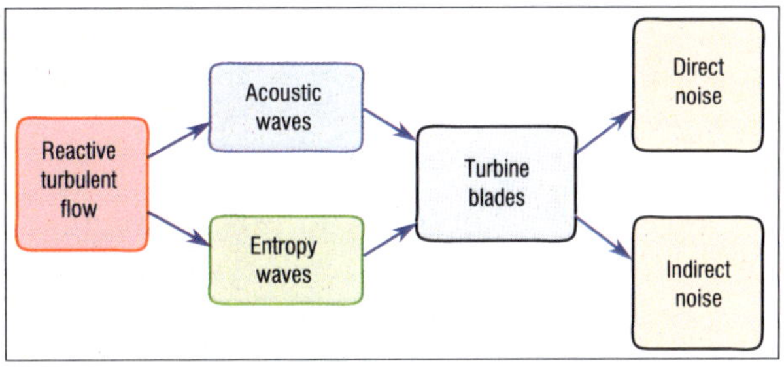

The two combustion noise mechanisms.

Indirect Combustion Noise

Any heat release generates entropy, as described by the equation:

$$\frac{\partial s}{\partial t}+\vec{u}.\nabla s=\frac{\dot{Q}}{\rho T} \text{ with } s=c_v \ln\left(\frac{p}{\rho^\gamma}\right)$$

where the usual definition of entropy is used. It is therefore straightforward to conclude that an unsteady heat release q' of this heat source (due to turbulence for example) generates a fluctuating entropy term, s' , given by a linearization:

$$\frac{\partial s'}{\partial t}+\vec{u}.\nabla s'=\frac{q'}{\rho T}-\left(\frac{p'}{p^\gamma}\vec{u}+\vec{u}'\right)\nabla s \text{ with } \frac{s'}{c_p}=\frac{p'}{\gamma p}-\frac{\rho'}{\rho},$$

as shown by Dowling et al. where the mean steady quantities are represented by the non-primed variables for simplicity. This equation shows first that entropy fluctuations are generated by both the flame fluctuations (q') and by acoustic fluctuations in regions with mean heat release (second term on the right hand side). At the same time, it shows that once generated, in the region outside the flame where no mean or fluctuating heat release exists, the right-hand-side of Eq. $\frac{\partial s'}{\partial t}+\vec{u}.\nabla s'=\frac{q'}{\rho T}-\left(\frac{p'}{p^\gamma}\vec{u}+\vec{u}'\right)\nabla s \text{ with } \frac{s'}{c_p}=\frac{p'}{\gamma p}-\frac{\rho'}{\rho}$, is zero and the entropy wave is just convected with the mean flow velocity ($\vec{u}$) u ρ when neglecting any flow mixing or diffusion. Marble and Candel showed that a one-way coupling between entropy and acoustics appeared when a flow gradient existed in a quasi-1D nozzle. Using the quasi-1D linearized Euler equations for the flow through the nozzle (neglecting heat release and diffusion),

$$\left[\frac{\partial}{\partial t}+u\frac{\partial}{\partial x}\right]\left(\frac{p'}{\gamma p}\right)+u\frac{\partial}{\partial x}\left(\frac{u'}{u}\right)=0$$

$$\left[\frac{\partial}{\partial t}+u\frac{\partial}{\partial x}\right]\left(\frac{u'}{u}\right)+\frac{c^2}{u}\frac{\partial}{\partial x}\left(\frac{p'}{\gamma p}\right)=\left[\frac{s'}{c_p}-2\frac{u'}{u}+(\gamma-1)\left(\frac{p'}{\gamma p}\right)\right]\frac{du}{dx}$$

$$\left[\frac{\partial}{\partial t}+u\frac{\partial}{\partial x}\right]\left(\frac{s'}{c_p}\right)=0$$

the coupling between the entropy and acoustics appears on the right hand side of eq. $\left[\frac{\partial}{\partial t}+u\frac{\partial}{\partial x}\right]\left(\frac{u'}{u}\right)+\frac{c^2}{u}\frac{\partial}{\partial x}\left(\frac{p'}{\gamma p}\right)=\left[\frac{s'}{c_p}-2\frac{u'}{u}+(\gamma-1)\left(\frac{p'}{\gamma p}\right)\right]\frac{du}{dx}$, where the entropy wave appears multiplied by the velocity gradient, du/dx. If the flow is homogeneous with no acceleration, the right hand side of eq. $\left[\frac{\partial}{\partial t}+u\frac{\partial}{\partial x}\right]\left(\frac{u'}{u}\right)+\frac{c^2}{u}\frac{\partial}{\partial x}\left(\frac{p'}{\gamma p}\right)=\left[\frac{s'}{c_p}-2\frac{u'}{u}+(\gamma-1)\left(\frac{p'}{\gamma p}\right)\right]\frac{du}{dx}$ is zero and the well-known result for acoustic propagation in non-zero but constant-Mach-number flows is recovered from eqs. $\left[\frac{\partial}{\partial t}+u\frac{\partial}{\partial x}\right]\left(\frac{p'}{\gamma p}\right)+u\frac{\partial}{\partial x}\left(\frac{u'}{u}\right)=0$ and $\left[\frac{\partial}{\partial t}+u\frac{\partial}{\partial x}\right]\left(\frac{u'}{u}\right)+\frac{c^2}{u}\frac{\partial}{\partial x}\left(\frac{p'}{\gamma p}\right)=\left[\frac{s'}{c_p}-2\frac{u'}{u}+(\gamma-1)\left(\frac{p'}{\gamma p}\right)\right]\frac{du}{dx}$. However, when considering a mean flow gradient, the coupling between convective and acoustic waves appears: the entropy wave acts as a source term in the acoustic equation, generating entropy or indirect combustion noise. This mechanism, which converts entropy waves into acoustic noise, has been extensively studied from an experimental , analytical and numerical point of view.

Direct Combustion Noise

As seen in the second part of eq. $\frac{\partial s'}{\partial t}+\vec{u}.\nabla s'=\frac{q'}{\rho T}-\left(\frac{p'}{p^{\gamma}}\vec{u}+\vec{u}'\right)\nabla s$ with $\frac{s'}{c_p}=\frac{p'}{\gamma p}-\frac{\rho'}{\rho}$, the entropy wave has an associated density fluctuation. In the near region of the flame, this density fluctuation is generated by the unsteady heat release and generates a velocity perturbation (acoustic wave) due to the mass conservation equation. This mechanism generates acoustic waves directly in the flame region as a monopole source, which propagate through the turbine stages, generating direct combustion noise. In free space, an acoustic analogy can be used to propagate this noise source. For example, when considering Phillips' analogy, which was extended to reactive flows by Chiu and Summerfield and Kotake, the wave equation reads,

$$\frac{D^2\pi}{Dt^2}-\frac{\partial}{\partial x_i}\left(c^2\frac{\partial\pi}{\partial x_i}\right)=\frac{D}{Dt}\left[\frac{(\gamma-1)}{\rho c^2}\dot{\omega}_T\right]$$

where $\pi=(1/\gamma)\ln\left(p/p_\infty\right)$ and $\dot{\omega}_T$ & is the heat release of the flame. Only this direct combustion noise contribution has been retained, for simplicity, to illustrate the monopole nature of combustion noise at low Mach numbers. Direct combustion noise has been studied analytically, combining numerical simulations and acoustic analogies and experimentally. These studies are mostly performed for free unconfined flames, a situation that is never observed in an engine.

The indirect mechanism can be of great importance in modern aero-engines due to the large accelerations of the flow in the turbine blades, as seen by. Leyko et al. showed that, for a 1D model combustor with a choked nozzle, indirect noise can be one order of magnitude larger than direct noise, depending on the mean Mach number in the combustion chamber. An order of magnitude can be useful here: sending a 10K temperature fluctuation through a choked nozzle can produce a pressure oscillation of up to 40 Pa at the outlet.

Both combustion noise generation mechanisms depend strongly on the propagation of waves through turbine stages: direct noise is shielded by the effect of the turbine blades and the successive accelerations and decelerations of the mean flow, while entropy noise is, on the other hand, generated by these accelerations and decelerations. For this reason, the study and the prediction of combustion noise has focused both on the generation of waves by the flame in the chamber and on the propagation of waves through non-uniform flows, which is typical of turbine stages.

Predicting Combustion Noise in Modern Aero-engines

The prediction of combustion noise through numerical methods is a challenging task. First, noise sources must be computed accurately in the combustion region, then propagated through turbine stages including the effects of velocity gradients where the entropy wave generates indirect noise and, finally, noise must be propagated in the far field. The hybrid method proposed here combines Large Eddy Simulations (LES) of the combustion chamber with analytical methods to resolve the propagation of waves through the turbine stages and is illustrated in figure.

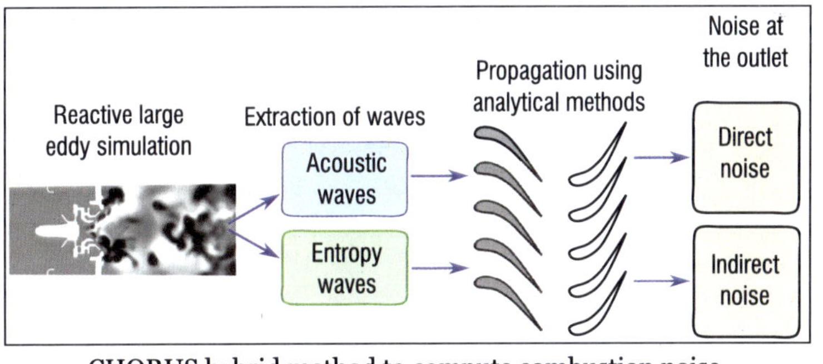

CHORUS hybrid method to compute combustion noise.

Since noise sources (in this case the flame dynamics) must be seized properly, RANS (Reynolds-Averaged Navier-Stokes) simulations are less suitable, since the unsteady features of the flow would have to be modeled. Indeed, the unsteadiness of the heat release is mainly caused by turbulence-flame interactions and acoustics-flame interactions, and thus a resolved compressible reactive LES is more suited. A proper LES simulation can resolve the large turbulent structures of the flow, which are responsible for the low-frequency wrinkling of the flame and therefore the unsteady heat release that generates noise. Small-scale turbulence is not resolved and must be modeled appropriately to consider its dissipation in the mean flow.

The flame model is of great importance also. The flame thickness is usually smaller than the grid size of any LES simulation and must be modeled or thickened artificially to be properly resolved in the computational mesh, thereby including an extra layer of modeling. The modeled flame must react properly to velocity, temperature and, in some cases, mixture fraction fluctuations in the flame front; therefore, the model can be a key element when computing combustion noise.

The flamelet formulation has been used for nonpremixed cases, while the G-equation has been used to model the flame in premixed combustion cases. A more general formulation is to use the thickened flame model, which has the advantage of dealing with both premixed and non-premixed flames. This is the case of modern aero-engines, where the flow is partially premixed and thus both premixed and diffusion (non-premixed) flames can be present. The idea behind this model is to artificially increase the flame thickness by multiplying the thermal diffusivity (D) by a factor F and dividing the pre-exponential constant (A) by the same factor. Since the flame thickness of premixed flames scales as $\delta \sim \sqrt{D/A}$ and the laminar flame speed as $s_1 \sim \sqrt{DA}$, the new flame would be F times thicker (in order to be resolved by the mesh), while conserving the original laminar flame speed. The drawback of this formulation is that the Damköhler number, $D_a \sim \tau_t / \tau_c$, is also decreased by a factor , meaning that vortices cannot wrinkle the thickened flame in the same way as they do with the real thin flame. The wrinkling of the flame by turbulent scales must be modeled through an efficiency function obtained from canonical DNS simulations. The advantage of this formulation, however, is first that a multi-species combustion with different non-unity Lewis numbers can be considered, since the flame front is resolved in the mesh and not only as a discontinuity of the flow and, at the same time, the level of modeling tends to zero when the mesh size is reduced, which is not the case for the flamelet and the G-equation models. This last point is of great importance since, when performing resolved LES, one expects the numerical solution to tends to the real one when the mesh size tends to zero.

Next, once the noise sources are computed using LES, to obtain the direct and indirect noises generated, the acoustic and entropy waves must be propagated from the outlet of the combustion chamber to the outlet of the turbine. This propagation through a non-homogeneous flow is responsible, on the one hand, for the attenuation of the acoustic waves and, on the other hand, for the generation of indirect noise through the successive acceleration and deceleration of entropy waves in each turbine stage. However, including the turbine in the LES numerical simulation is not feasible using state-of-the-art computational codes and resources, and some kind of acoustic analogy or modeling should be used to account for the propagation and generation of indirect noise through the turbine. Acoustic analogies such as that of Lighthill are widely used in the aeroacoustics community and easily applied with a low computational cost. When writing the complete acoustic analogy for reacting flows:

$$
\begin{aligned}
&\frac{1}{c_\infty}\frac{\partial^2 p}{\partial t^2}-\frac{\partial 2p}{\partial x_i^2}=\frac{\partial 2}{\partial x_i x_j}(\rho u_i u_j-\tau_{ij})\\
&+\frac{\partial}{\partial t}\left[\begin{array}{l}\dfrac{\rho_\infty}{\rho}\dfrac{(\gamma-1)}{c^2}\left(\omega_T^{\bullet}+\sum_k h_k\dfrac{\partial J_k}{\partial x_k}-\dfrac{\partial q_i}{\partial x_i}+\tau_{ij}\dfrac{\partial u_i}{\partial x_j}+\dot{Q}\right)\\ +\rho_\infty\dfrac{D}{Dt}\mathrm{In}\, r\end{array}\right]\\
&+\frac{1}{c_\infty}\frac{\partial}{\partial t}\left[\left(1-\frac{\rho_\infty c_\infty^2}{\rho c^2}\right)\frac{Dp}{Dt}-\frac{p-p_\infty}{\rho}\frac{D\rho}{Dt}\right]\\
&+\frac{\partial^2}{\partial x_j t}(\rho_e u_j)
\end{aligned}
$$

The indirect combustion noise source (last line) appears on the right hand side of the wave equation, in terms of the excess density $\rho_e = (\rho - \rho_\infty) - (p - p_\infty)/c_\infty$, i.e, the density fluctuation that arises over and above isentropic acoustic contributions and associated with entropy fluctuations. A first drawback, however, is that the mean flow is considered at rest in the propagation region and therefore indirect noise, generated where velocity gradients are strong in the mean flow, cannot be predicted using this method and must be given in advance. Moreover, an acoustic analogy requires the medium to be broken down into a source region and a propagation zone, which can hardly be done in the indirect noise context. Assuming that the sound can be split into generation and propagation, the source term given by the excess density (right hand side of Eq $\frac{1}{c_\infty}\frac{\partial^2 p}{\partial t^2} - \frac{\partial 2p}{\partial x_i^2} = \frac{\partial 2}{\partial x_i x_j} + \frac{\partial^2}{\partial x_j t}(\rho_e u_j)$ must be part of the sources solved by LES to be propagated by the wave operator (left hand side of Eq $\frac{1}{c_\infty}\frac{\partial^2 p}{\partial t^2} - \frac{\partial 2p}{\partial x_i^2} = \frac{\partial 2}{\partial x_i x_j} \ldots\ldots + \frac{\partial^2}{\partial x_j t}(\rho_e u_j)$), which remains a daunting task. The alternative is to solve analytically (under several assumptions) the propagation of waves through the non-homogeneous flow. Marble and Candel first solved the propagation of waves through a quasi-1D subsonic and supersonic nozzle analytically using the compact assumption, where the wavelength of the waves is small compared to the nozzle transversal length (and therefore the result is valid in the low frequency range only). Thus, the linearized Euler equations $\left[\frac{\partial}{\partial t} + u\frac{\partial}{\partial x}\right]\left(\frac{p'}{\gamma p}\right) + u\frac{\partial}{\partial x}\left(\frac{u'}{u}\right) = 0$ - $\left[\frac{\partial}{\partial t} + u\frac{\partial}{\partial x}\right]\left(\frac{u'}{u}\right) + \frac{c^2}{u}\frac{\partial}{\partial x}\left(\frac{p'}{\gamma p}\right) = \left[\frac{s'}{c_p} - 2\frac{u'}{u} + (\gamma - 1)\left(\frac{p'}{\gamma p}\right)\right]\frac{du}{dx}$, $\left[\frac{\partial}{\partial t} + u\frac{\partial}{\partial x}\right]\left(\frac{s'}{c_p}\right) = 0$ are assumed to be quasi-steady, giving three jump conditions between the inlet (i) and the outlet (o) corresponding to the mass, total temperature and entropy fluctuations, namely

$$\left(\frac{\dot{m}'}{\dot{m}}\right)_i = \left(\frac{\dot{m}'}{\dot{m}}\right)_o$$

$$\left(\frac{T_t'}{T_t}\right)_i = \left(\frac{T'}{T_t}\right)_o$$

$$\left(\frac{s'}{c_p}\right)_i = \left(\frac{s'}{c_p}\right)_o$$

These matching conditions are written as a function of the dimensionless upstream and downstream acoustic propagating waves w+, w– and entropy wave ws defined in Eq ($w^+ = \frac{p'}{\gamma p} + \frac{u'}{\gamma c}$, $w^- = \frac{p'}{\gamma p} + \frac{u'}{\gamma c}$ and $w^s = \frac{s'}{C_p}$) and solved analytically. Fully analytical solutions exist also when considering non-compact frequencies for a specific nozzle geometry, for any nozzle geometry using an asymptotic expansion up to first order and, more recently, using the Magnus expansion for any frequency and any nozzle geometry. These methods, however, are based on a quasi-1D formulation, neglecting the real 3D flow of the turbine and, in particular, the flow deviation imposed by the stator and rotor blades.

$$w^{+} = \frac{p'}{\gamma p} + \frac{u'}{\gamma c}$$

$$w^{-} = \frac{p'}{\gamma p} + \frac{u'}{\gamma c}$$

$$w^{s} = \frac{s'}{C_p}$$

Cumpsty and Marble used the same compact assumption as Marble and Candel to write the jump conditions of the acoustic and entropy waves, as well as vorticity waves w^{v},

$$w^{v} = \frac{\xi'}{\omega}$$

where ξ' is a vorticity fluctuation, through a two-dimensional stator blade row combining Marble's equation's with the Kutta condition at the outlet of the row. This formulation is still based on the same compact assumption and is therefore only valid for low frequencies, but considers the 2D azimuthal component of the flow though neglecting radial effects, which are small below the radial cut-off frequency, which is typically of the order of 103 −104 Hz. The model includes the effects of flow deviation, vorticity, the different azimuthal modes present in the combustion chamber and their cut-off frequencies, being therefore more suitable for the propagation and generation of combustion noise in turbine stages.

Finally, once the acoustic waves are obtained at the outlet of the aeroengine (direct and indirect), an acoustic analogy or a linearized 3D computation is used to obtain the far field combustion noise spectrum. The complete hybrid computation chain (named CHORUS) is illustrated in figure. The compressible unstructured reactive LES solver AVBP [42, 55] is used to compute the noise sources in the complex geometry of an actual combustion chamber. From this computation, the outgoing dimensionless waves defined above are extracted at the outlet of the combustion chamber and propagated through the blade rows using analytical method. The waves at the outlet of the turbine can be finally propagated to the far-field using the acoustic solver AVSP-F, which solves the 3D linearized Euler equations in a quiescent flow in the frequency domain.

This hybrid method has been successfully applied in several stateof-the-art industrial aero-engines. The results of one of such simulations are presented here as an example. The LES simulations of the combustion chamber are performed on a single sector, using the unstructured compressible reactive code AVBP. The waves at the outlet of the combustion chamber (acoustic, entropy and vorticity) are plotted in figure. It can be seen that the acoustic waves are mainly longitudinal. It should be noted that, since the LES simulation is performed on a single sector, the low-order circumferential modes cannot be computed and only the plane mode will be taken into account. The circumferential mode in figure corresponds to the first multiple of the number of sectors (m =18).

Using the analytical methods of Cumpsty and Marble, CHORUS is able to propagate these waves through the turbine stages and compute, on the one hand, direct combustion noise generated by acoustic waves and, on the other hand, the contribution from entropy and vorticity waves in indirect combustion noise. Below 8400 Hz, the azimuthal mode, visible at the outlet of the chamber in figure, is evanescent. As a consequence, in this computation, the plane mode is the only contributor to noise generation through the turbine. Noise terms are plotted in figure in dB, showing that vorticity noise represents the weakest contribution and that indirect noise is stronger than direct combustion noise, though the latter is still significant and cannot be neglected.

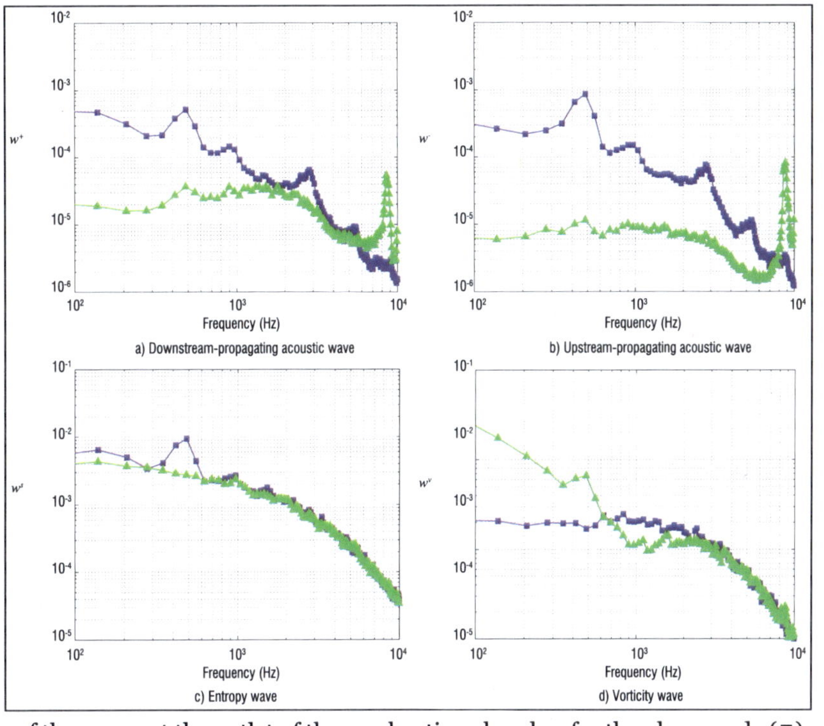

Spectra of the waves at the outlet of the combustion chamber for the plane mode (■) and for the first circumferential mode of the sector, m =18 (Δ).

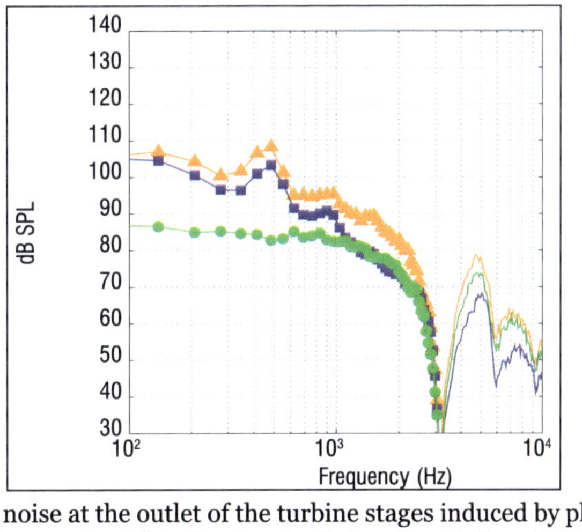

Combustion noise at the outlet of the turbine stages induced by plane waves. Indirect noise (Δ), direct noise (■) and vorticity noise (•).

Experiments on Combustion Noise

Strahle, Giammar and Putnam give reviews about experiments performed before the 1970s on combustion noise, mostly focused on direct combustion noise. Recent experiments have focused on indirect combustion noise, on the one hand to identify whether this mechanism is significant and, on the other hand, to assess the existing analytical theories for the prediction of this type of noise.

NOISE SOURCES IN GAS TURBINES

Engine noise is generated by all engine sections: the fan, the compressor, the combustor, the turbine and the exhaust also referred to as the jet. The main focus of this factsheet is on noise generated by the fan and the exhaust. Most noise reduction techniques are applied on these two modules, since they generate most of the noise.

Turbofan engine with major noise components and their directions shows a graphical representation of noise emission.

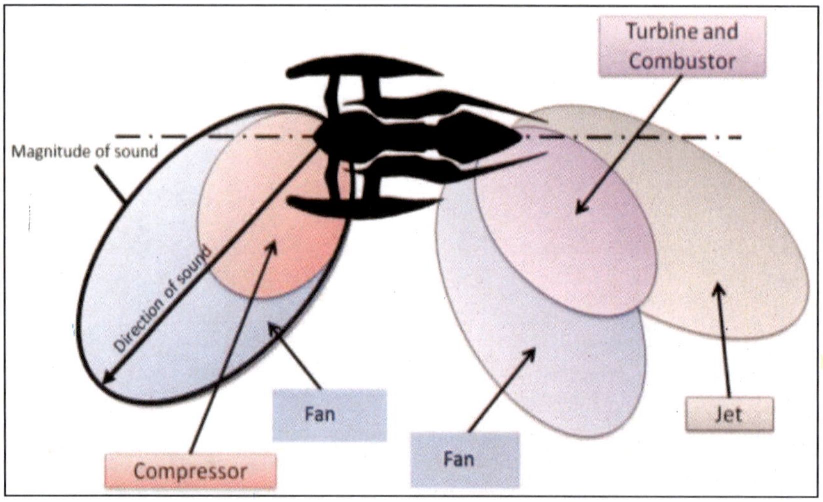

Turbofan engine with major noise components and their directions.

Fan Noise

Sound is produced by distortion such as pressure fluctuation – in an elastic medium. A fan system will produce sounds with tonal characteristics and broadband characteristics.

- Tonal sound: A sound or noise recognizable by its regularity. A simple or pure tone a musical note, for example has one frequency.
- Broadband sound: Noise whose energy is distributed over a wide section of the audible range, such as noise coming from wind.

Because these two sound types are generated differently, a separate discussion will be provided for each of them.

Tonal Noise

Tonal noise is generated when there is an interactive effect between airflow distortions in the path of a rotating blade. Most frequently, these interactions involve pressure fields or turbulent-wake disturbances.

A discrete tone, propagated to the far-field, is generated by an almost identical pressure field. This pressure field is associated with each blade on a single rotating stage. The propagation to the far-field occurs when a signal is recognized as a sound-wave. Usually this sound-wave is a wavelength to where the sound pressure and the particle velocity are in phase.

At subsonic conditions, when no shock pattern exists, the cyclic pressure field and wake interactions between rotating and stationary stages are the source of discrete tones. If the stages are close together, there is generally an intense pressure-field interaction and a very strong tone generation.

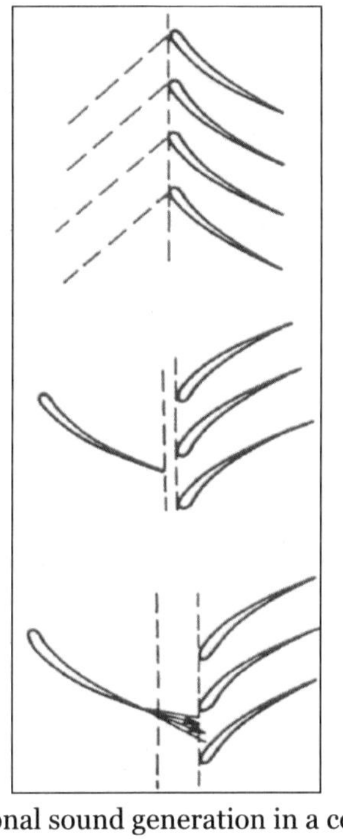

Mechanisms of tonal sound generation in a compressor stage.

Broadband Noise

Broadband noise results from the propagation of sound produced near the surface of the blade as a result of pressure fluctuations associated with turbulent flow in the vicinity.

In the first stage of the compressor – or the fan in a turbofan engine – broadband noise is generated by the movement of the tip of the rotating blade within the turbulent boundary layer close to the wall of the inlet duct.

Turbulence in the wakes shed by fan blades is also an important source of broadband noise, particularly for fan blades with large surface areas. It plays a significant role in the generation of broadband noise in the downstream stages of the turbo machine. In a modern single-stage fan, these are the outlet stators in the bypass duct and the multistage core compressor.

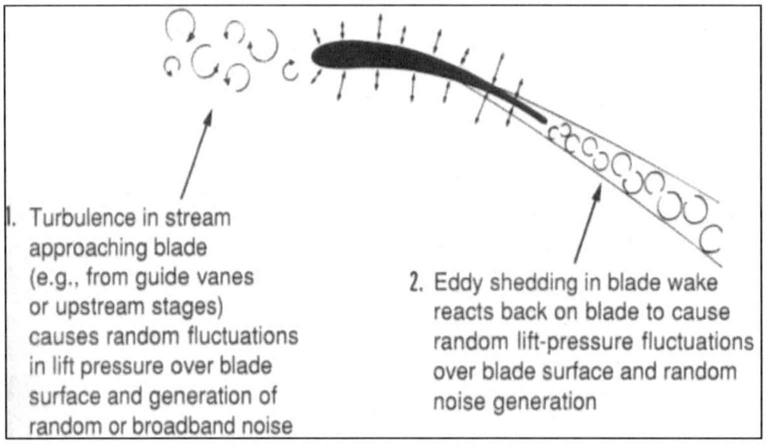

Generation of broadband noise.

NOISE REDUCTION TECHNIQUES IN GAS TURBINES

The techniques used for noise reduction in gas turbines reduce fan noise and jet noise. The new generation gas turbines mainly cope with broadband noise rather than tones.

Fan Noise

Fan noise is lowered by reducing the rotational tip speed of the rotor blades or reducing the fan pressure ratio. This causes the engine diameter to increase in order to achieve the same thrust production. The best fan tip speed is just below the speed of sound. Once this has been achieved in an engine design, the fan pressure ratio becomes the controlling factor for reducing (broadband) noise. Techniques for reducing fan noise are: the UHBR engine, scarf inlets, active noise control, forward swept fans, swept and leaned stators, variable area fan nozzles, fan trailing edge blowing, soft stator vanes, and acoustic treatment.

Ultra-high Bypass Ratio (UHBR)

The ultra-high bypass ratio engine is an engine with an ultra-high bypass, as depicted in figure Compared with a bypass ratio engine, the nacelles of an UHBR engine are considerably reduced in length and volume. By decreasing the bypass nacelles of the turbofan, noise is lowered through a reduction in the rotational tip speed of the fan blades and a reduction in the fan pressure ratio. By decreasing the diameter, the UHBR solution is the most effective and easiest way to reduce noise. The other main advantage of the UHBR engine is that it increases sustainability in an efficient way.

Scarf Inlets

Scarf inlets reduce fan noise by redirecting the forward-radiated sound – both tones and broadband noise – away from the community (upwards), as depicted in figure. Scarf inlets also have aerodynamic advantages.

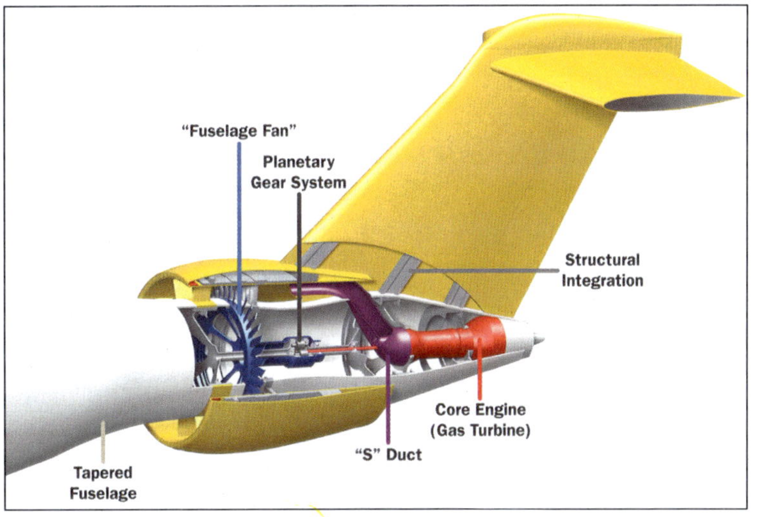

Ultra-high bypass ratio engine including scarf inlet.

Active Noise Control

Active noise control is used in aviation gas turbine fans, called the Active Noise Control Fan (ANCF). An actuator in the fan duct walls and inside the stator vanes captures the tones of a fan and directly send a 180 degree shifted tone back (anti-noise) to cancel out the noise source.

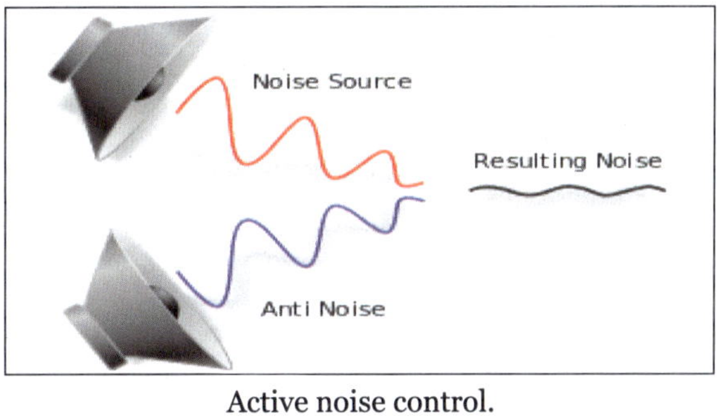

Active noise control.

Forward Swept Fans

Forward swept fans, also called forward swept fan tip blades mainly reduce aerodynamic losses associated with shocks and improve stall margin. By reducing the opportunity of shocks, noises associated with these shocks are eliminated by delaying the onset of multiple pure tones.

Forward swept fans.

Swept and Leaned Stators

Swept and leaned stators are depicted in figure. They reduce fan noise by increasing the phase changes from root to tip – and the effective distance from the fan to the stator vanes (both producing the sound). Increasing the effective distance and phase changes creates less turbulence. Swept and leaned stators reduce broadband noise as well as tones. Unfortunately swept and leaned stator blades have additional aerodynamic performance losses, resulting in a trade-off between noise reduction and performance.

Swept and leaned stators.

Variable Area Fan Nozzles

A variable area fan nozzle is a moving mechanism that can increase and decrease the area of the fan nozzle. Controlling the incidence angle of the flow reduces noise coming from the fan. If the fan nozzle area increases, then the fan noise coming from the back of the engine decreases. Variable area fan nozzles can also increase and decrease thrust.

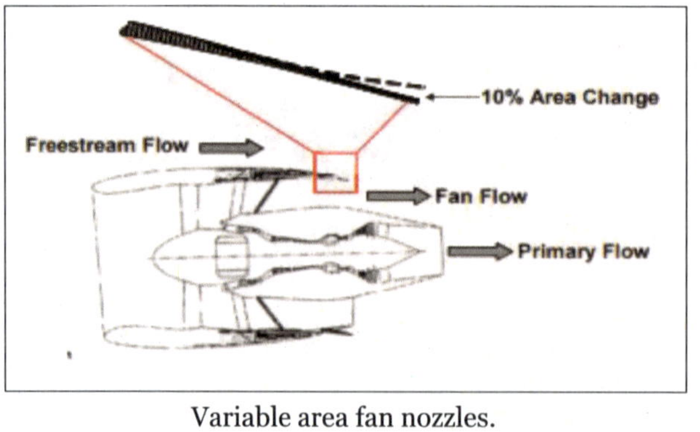

Variable area fan nozzles.

Fan Trailing Edge Blowing

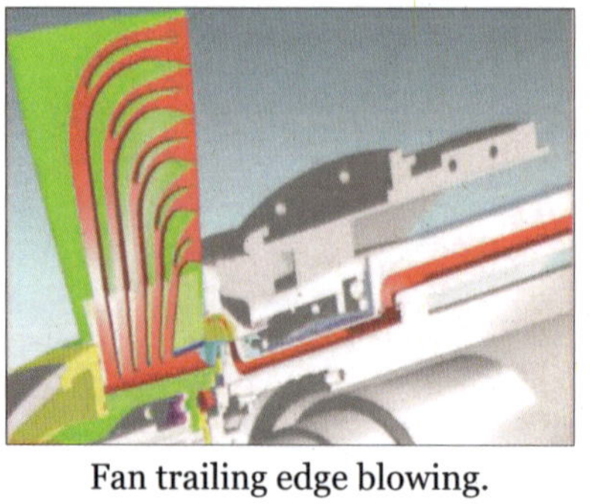

Fan trailing edge blowing.

Fan noise can also be reduced by ejecting air at the trailing edge of the fan blades. Specially-designed passages pump air from the hub of the fan to the trailing edge of the fan, as depicted in figure. Air is pumped into the engine to smoothen the air flow. By smoothening the air flow, fan trailing edge blowing reduces the wake defect and thereby tone noises. Fan trailing edge blowing also reduces the turbulence intensity of the fan, which reduces broadband noise.

Soft Stator Vanes

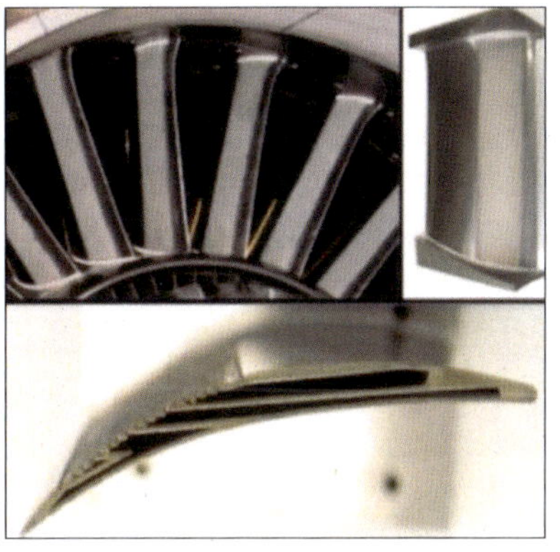

Soft stator vanes.

Soft stator vanes reduce tone noise coming from the fan. They are vanes on the trailing edge of a porous surface (also referred to as soft), with hollow chambers, as depicted in figure. Each chamber is tuned in a specific way to damp out a particular frequency. Soft stator vanes reduce the unsteady pressure response on the stator surface and absorb energy that would otherwise become sound radiating from the stator. Soft stator vane design can be tailored for specific frequency ranges.

Acoustic Treatment

Acoustic treatment is accomplished by putting acoustic liners in the inlet, aft fan ducts and the tip of the rotor. These liners are mostly honeycomb structures with porous or felt metal face sheets. These sheets reduce tone noise around a desired target frequency. Honeycomb structures reduce fan noise by wearing out the tones.

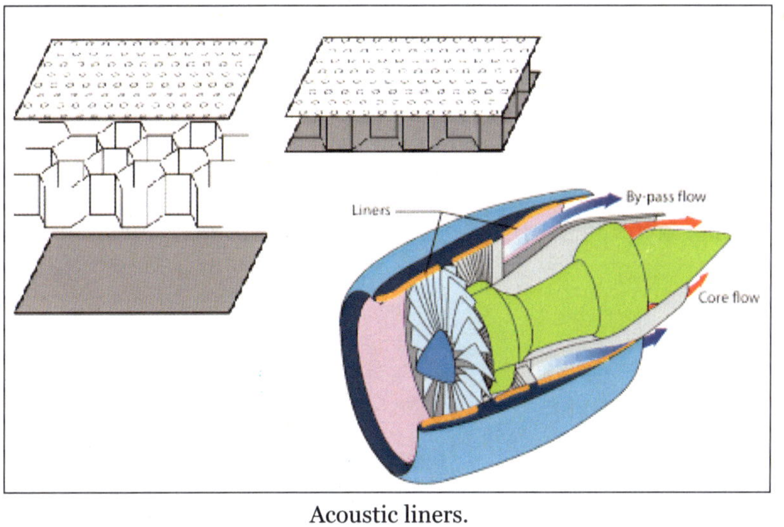

Acoustic liners.

JET NOISE

Jet noise is the field that focuses on the noise generation caused by high-velocity jets and the turbulent eddies generated by shearing flow. Such noise is known as broadband noise and extends well beyond the range of human hearing (100 kHz and higher). Jet noise is also responsible for some of the loudest sounds ever produced by mankind.

Sources of Jet Noise

The primary sources of jet noise for a high-speed air jet (meaning when the exhaust velocity exceeds about 100 m/s) are "jet mixing noise" and, for supersonic flow, shock associated noise. Also, acoustic sources within the "jet pipe" also contribute to the noise, mainly at lower speeds, which include combustion noise and sounds produced by interactions of a turbulent stream with fans, compressors, and turbine systems.

The jet mixing sound is created by the turbulent mixing of a jet with the ambient fluid, in most cases, air. The mixing initially occurs in an annular shear layer, which grows with the length of the nozzle. The mixing region generally fills the entire jet at four or five diameters from the nozzle. The high-frequency components of the sound are mainly stationed close to the nozzle, where the dimensions of the turbulence eddies are small. Further down the jet, where the eddy size is similar to the jet diameter, is where lower frequency begins.

In Supersonic, or "choked" jets there are cells through which the flow continuously expands and contracts. Several of these "shock cells" can be seen extending up to ten jet diameters from the nozzle and are responsible for two additional components of jet noise, "screech tones" and broadband "shock associated noises". Screech is produced by a feedback mechanism in which a disturbance convecting in the shear layer generates sound as it traverses the standing system of shock waves in the jet. Even though screech is a side effect of the jet's flight, it can be suppressed by an appropriate design for a nozzle.

Aircraft noise is also sometimes called "jet noise" when emanating from jet aircraft, regardless of the mechanism of noise production.

JET NOISE SUPPRESSION

Jet Plume Aerodynamics

A prediction of the jet plume aerodynamics is required to provide the strength of the noise sources. The method selected is an extension of Reichardt's theory, which basically synthesizes the complex flows from nozzles of arbitrary geometry by superposition of a suitable distribution of elemental round jet flows.

Reichardt's theory is a semiempirical one, based on extensive experimental observations that the axial momentum flux profiles are bell shaped or Gaussian in the fully developed similarity region (far downstream of the exit plan) of a jet.

From these observations a hypothesis for the relation between axial and transverse momentum flux was formulated, and this hypothesis yields a governing equation for the axial momentum flux. For the similarity region of a circular jet with nozzle area A_j and exit velocity V_j, the governing equation and solution are as follows:

$$\frac{\partial}{\partial x}\left\langle \rho u^2 \right\rangle = \frac{\lambda(x)}{r}\frac{\partial}{\partial r}\left(r\frac{\partial}{\partial r}\left\langle \rho u^2 \right\rangle \right)$$

$$\left\langle \rho u^2 \right\rangle = \rho_j V_j^2 \frac{A_j}{\pi b_m^2}\exp[-(r/b_m)^2]$$

Where,

$$\lambda(x) = 1/2 b_m \, db_m / dx$$

u is the axial velocity component, ρ is density, the angle brackets ( ) represent the statistical time average, and $b_m(x)$ is the width of the axial momentum mixing region, taken to be proportional to the axial distance from the nozzle exit plane:

$$b_m(x) = C_m x$$

The jet spreading rate C_m becomes a key parameter in the theory and is determined experimentally. The coordinate system is shown in figure.

Because equation $\frac{\partial}{\partial x}\left\langle \rho u^2 \right\rangle = \frac{\lambda(x)}{r}\frac{\partial}{\partial r}\left(r\frac{\partial}{\partial r}\left\langle \rho u^2 \right\rangle \right)$ is linear, the summation of elemental solutions (eq. $\left\langle \rho u^2 \right\rangle = \rho_j V_j^2 \frac{A_j}{\pi b_m^2}\exp[-(r/b_m)^2]$) is also a solution. This unique feature of Reichardt's theory allows the construction of quite complex jet flows with relatively simple mathematics. Although more rigorous (but containing just as much empiricism, albeit in different forms) theories are available for simple jets (circular and planar), there is no other technique available which offers the capability for modeling jet flows typical of aircraft engine suppressor nozzles such as multitube, lobe, and chute nozzles.

Consider a distribution of elemental jets issuing parallel to the X-axis. The jet exit areas lie in the x = o plane. Each elemental jet has an exit area Aj = σ, dσ, da located at (σ, α, o), as shown in figure. The axial momentum flux at a downstream point (r, θ, x) due to the elemental jet exhausting at (σ, α, o) is given by (from eq. $\left\langle \rho u^2 \right\rangle = \rho_j V_j^2 \frac{A_j}{\pi b_m^2}\exp[-(r/b_m)^2]$)

$$d\left\langle \rho u^2 \right\rangle = \rho_j V_j^2 (\sigma \, d\sigma \, d\alpha / \pi b_m^2)\exp[-\xi/b_m)^2]$$

Where,

$$\xi^2 = r^2 + \sigma^2 - 2r\sigma\cos(\theta - \alpha)$$

Integrating eq. $u\left\langle \rho\ ^2 \right\rangle = \rho_j V_j^2 (\sigma \, d\sigma \, d\alpha / \pi b_m^2)\exp[-\xi / b_m)^2$ results in the following solution:

$$\left\langle \rho u^2 \right\rangle(r,\theta,x) = \frac{1}{\pi b_m^2}\iint (\rho_j V_j^2)\exp[-(\xi/b_m)^2]\sigma \; d\sigma \, d\alpha$$

From the distribution of $\rho_j V_j^2$ in the exit plane, the local value of $\langle \rho u^2 \rangle$ at any point (r, θ, x) can be found from equation $\langle \rho u^2 \rangle (r,\theta,x) = \frac{1}{\pi b_m^2} \iint (\rho_j V_j^2) \exp[-(\xi/b_m)^2]\sigma\, d\sigma\, d\alpha$ by standard numerical integration. If we assume that the jet plume stagnation enthalpy flux H diffuses in the same manner as axial momentum, an analogous expression for stagnation enthalpy flux $\langle \rho\, uH \rangle$ can be derived:

$$\langle \rho u H \rangle (r,\theta,x) = \frac{1}{\pi b_h^2} \iint (\rho_j V_j H_j) \exp[-(\xi/b_h)^2]\sigma\, d\sigma\, d\alpha$$

where b_h is the width of the thermal shear layer, taken to be proportional to x:

$$b_h = C_h x$$

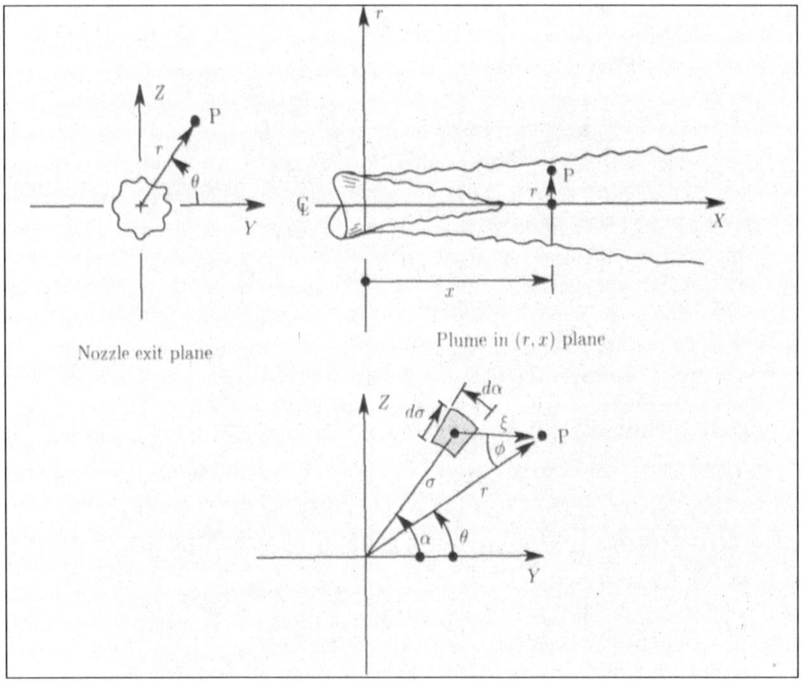

Nomenclature for elemental jet at ($\sigma, a, 0$) and field point P Jet flow coordinate system and nomenclature.

for constant C_h. The stagnation enthalpy is defined as $H = c_p T_j + (u^2/2) - c_p T_o$ (where c_p is specific heat at constant pressure, T_j is jet temperature, and T_o is ambient temperature), and the thermal layer spreading rate C_h also must be obtained experimentally. If we assume that the jet mixing occurs at constant static pressure equal to the ambient value, the solutions for $\langle \rho u^2 \rangle$ and $\langle \rho u H \rangle$ given by equations $\langle \rho u^2 \rangle (r,\theta,x) = \frac{1}{\pi b_m^2} \iint (\rho_j V_j^2) \exp[-(\xi/b_m)^2]\sigma\, d\sigma\, d\alpha$ and $\langle \rho u H \rangle (r,\theta,x) = \frac{1}{\pi b_h^2} \iint (\rho_j V_j H_j) \exp[-(\xi/b_h)^2]\sigma\, d\sigma\, d\alpha$ are sufficient to determine the distributions of mean axial velocity u and temperature T_j throughout the jet plume.

In addition to the jet plume mean-flow properties, the turbulent Reynolds stress which is assumed to be proportional to the transverse momentum flux, also can be obtained. Reichardt's hypothesis (from which eq. $\frac{\partial}{\partial x}\left\langle \rho u^2 \right\rangle = \frac{\lambda(x)}{r}\frac{\partial}{\partial r}\left(r\frac{\partial}{\partial r}\left\langle \rho u^2 \right\rangle \right)$ evolved) states that the transverse momentum flux is proportional to the transverse gradient of the axial momentum flux, the proportionality factor being A(x). For a simple circular jet (eqs. $\left\langle \rho u^2 \right\rangle = \rho_j V_j^2 \frac{A_j}{\pi b_m^2} \exp[-(r/b_m)^2]$, $\lambda(x) = 1/2 b_m \, db_m / dx$, $b_m(x) = C_m x$), the Reynolds stress τ is given by:

$$\tau = -\left\langle \rho u'v' \right\rangle \approx -\lambda \frac{\partial}{\partial \tau}\left\langle \rho u^2 \right\rangle = C_m \rho_j V_j^2 \frac{A_j}{\pi b_m^2}\frac{\tau}{b_m} \exp[-(r/b_m)^2]$$

where u' and v' are the axial and radial components of turbulent fluctuation velocity. For an elemental jet exhausting at (σ, α, 0) the shear stress at (r, θ, x) lies along a line connecting (σ, α, 0) and the projection of (r, θ, x) onto the x = 0 plane. This vector is at an angle 0 to the coordinate direction r. The radial component of the shear stress dτ at point (r, θ, x) resulting from an elemental jet exhausting at (σ, α, 0) is then $d\tau_r$ = dτ cos ∅. Similarly, the azimuthal component is $d\tau_r$ = dθ sin ∅. If we perform the same summation and limiting process over all elemental jets, the total shear stress at (r, θ, x) is:

$$\tau = \left(\tau_r^2 + \tau_\theta^2 \right)^{1/2}$$

Where,

$$\tau_r(r,\theta,x) = \frac{C_m}{\pi b_m^2} \iint \rho_j V_j^2 (\xi / b_m) \exp[-(\xi / b_m)^2](\cos\phi)\sigma \; d\sigma \, d\alpha$$

The azimuthal shear stress τθ(r, θ, x) is given by a similar expression with cos ∅ replaced by sin ∅. The distance is again given by the expression from equation $d\left\langle \rho u^2 \right\rangle = \rho_j V_j^2 (\sigma \, d\sigma \, d\alpha / \pi b_m^2)$ $\exp[-\xi / b_m)^2]$, and the angle ∅ is given by,

$$\xi \cos\phi = r - \sigma\cos(\theta - \alpha)$$

The Eight above mentioned equations provide the basic expressions for computation of the jet plume flow parameters T_j,u, and T for a nozzle of arbitrary exit cross section and exit distribution of velocity and temperature. For axisymmetric nozzles, $\tau = \tau_r$ and $\tau_\theta = 0$. The basic limiting assumptions underlying this aerodynamic model for the jet plume characteristics are (1) the jet plume mixing occurs at constant static pressure, equal to the ambient value, and (2) the flow is primarily axial, with all nozzle exit elements in the same plane (x = 0).

Intrinsic Source Intensity Mixing-noise Spectrum

The aerodynamic characteristics of the jet plume provide the information required to evaluate the acoustic intensity spectrum in the absence of convection and acoustic mean-flow interaction effects.

This represents the sound spectrum which would be heard if the turbulent eddies generating the sound had negligible convection speed relative to the observer and if the velocity and temperature gradients encountered by the sound as it propagated through the jet plume itself had no effect on the sound radiation. As is demonstrated, these effects are indeed powerful for high-velocity jets. The postulation herein is that the basic source strength spectrum can be developed, and the convection and acoustic—mean-flow interaction effects can then be added to this basic (intrinsic) sound spectrum.

The jet plume is first subdivided into elemental "eddy" volumes, each having its own source strength, spectrum, and flow shrouding, as illustrated in figure. The jet noise generation is represented by a collection of convecting and uncorrelated quadrupole sources, each radiating noise with an intensity spectrum directly related to the local flow properties. The net radiation from each eddy is a function of the flow environment of that eddy. Each eddy volume contains a mix of quadrupoles of various orientations. By employing a model of homogeneous, isotropic turbulence in the moving-eddy reference frame and taking the azimuthal average of the resulting sound field, Ribner found that a weighted combination of the various quadrupole types contributes to the net far-field radiation. The amplitude of these quadrupole types is of the form,

$$dI(\omega) = \frac{\rho_o l^3}{c_0^5 R^2}(u')^4 \omega^4 H(\mu)\, dV$$

where $dI(\omega)$ is the acoustic intensity per elemental jet volume dV, ρo is the ambient density, R is the source-to-observer distance, c o is the ambient speed of sound, U' is the local turbulence intensity, H(μ) is the Fourier transform of the moving frame space-time cross correlation of u', and μ is the ratio of emitted frequency ω to characteristic frequency ωo. Equation $dI(\omega) = \frac{\rho_o l^3}{c_0^5 R^2}(u')^4 \omega^4 H(\mu)\, dV$ is used to calculate the mixing noise amplitude and frequency content for each volume element in the jet. The characteristic turbulence frequency and length scale for each eddy volume are determined from the local mean-flow velocity, temperature, and shear stress.

$$\left.\begin{aligned} \omega_o &\approx \partial U / \partial r \\ l &\approx u' / \omega_o \end{aligned}\right\}$$

where l is the characteristic turbulent eddy size and the turbulence intensity is obtained from the shear stress as follows:

$$u' \approx \sqrt{\tau / \rho}$$

Acoustic Mean-flow Interaction Model

The equation which describes the propagation of sound emitted by the turbulence in a jet is as follows:

$$\frac{1}{c^2} D_*^3 p - D_*(\Delta p) \frac{d}{dr}(\log c^2) D_* \left(\frac{\partial p}{\partial r}\right) + 2\frac{\partial u}{\partial r}\frac{\partial^2 p}{\partial \times \partial r} = S$$

Where,

$$D_* = \frac{\partial}{\partial t} + U\frac{\partial}{\partial x} \quad \text{and} \quad S = \rho D_* [\nabla \cdot \nabla \cdot (u' \cdot u')]$$

In equations $\frac{1}{c^2} D_*^3 p - D_*(\Delta p) \frac{d}{dr}(\log c^2) D_* \left(\frac{\partial p}{\partial r}\right) + 2\frac{\partial u}{\partial r}\frac{\partial^2 p}{\partial \times \partial r} = S$ and $D_* = \frac{\partial}{\partial t} + U\frac{\partial}{\partial x}$ and $S = \rho D_* [\nabla \cdot \nabla \cdot (u' \cdot u')]$, U = U(r), c = c(r), and p = p(r) are the azimuthally averaged mean-flow axial jet velocity, speed of sound, and density, respectively. The symbol Δ is the Laplacian operator, t is time, and u' is essentially the turbulent velocity fluctuation. Roughly speaking, S is the noise source strength which drives the acoustic pressure fluctuations p. Equation $\frac{1}{c^2} D_*^3 p - D_*(\Delta p) \frac{d}{dr}(\log c^2) D_* \left(\frac{\partial p}{\partial r}\right) + 2\frac{\partial u}{\partial r}\frac{\partial^2 p}{\partial \times \partial r} = S$ is a third-order wave equation for the variable p, and this equation explicitly displays the influence of mean velocity U(r) and temperature (through speed of sound c(r)) profiles on the propagation of noise.

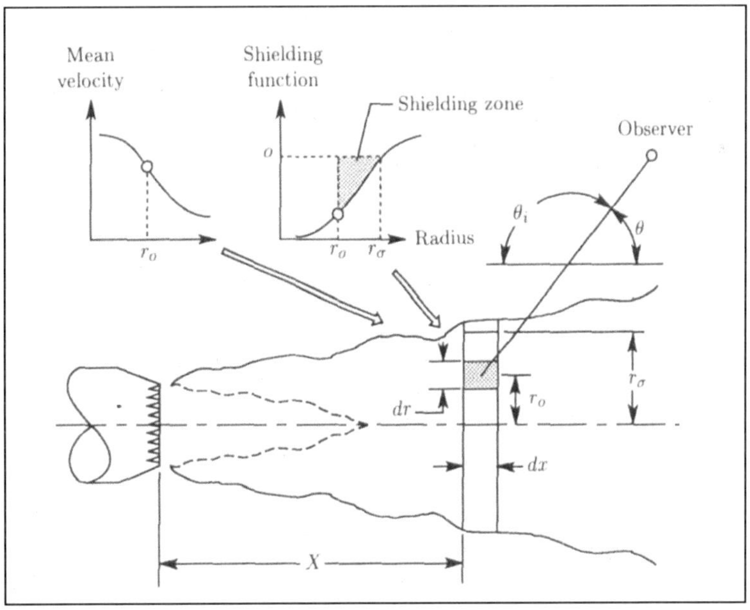

Generalized volume element model.

To solve equation $\frac{1}{c^2} D_*^3 p - D_*(\Delta p) \frac{d}{dr}(\log c^2) D_* \left(\frac{\partial p}{\partial r}\right) + 2\frac{\partial u}{\partial r}\frac{\partial^2 p}{\partial \times \partial r} = S$, the profiles for U(r), c(r), p(r), and S(r) must first be prescribed. The Green's function solution for equation $\frac{1}{c^2} D_*^3 p - D_*(\Delta p) \frac{d}{dr}(\log c^2) D_* \left(\frac{\partial p}{\partial r}\right) + 2\frac{\partial u}{\partial r}\frac{\partial^2 p}{\partial \times \partial r} = S$, when convoluted with the source term S, provides the solution to Lilley's equation. From the Green's function, solutions for higher order singularities (dipoles and quadrupoles) can be obtained by differentiating the source solution with respect to source coordinates. The right-hand side of equation

$\frac{1}{c^2} D_*^3 p - D_*(\Delta p) \frac{d}{dr} (\log c^2) D_* \left(\frac{\partial p}{\partial r} \right) + 2 \frac{\partial u}{\partial r} \frac{\partial^2 p}{\partial \times \partial r} = S$, S(r), represents a mix of quadrupoles of various orientations. The various quadrupole contributions are added, but with each quadrupole type having its own solution form representing the combined effects of convection and acoustic-mean-flow interaction.

The combined convection and interaction effects can be thought of as a multiplier of the basic source intensity spectrum given by equation $dI(\omega) = \frac{\rho_o l^3}{c_0^5 R^2} (u')^4 \omega^4 H(\mu)\, dV$, and this multiplier is a function of the local mean-flow properties and their radial gradients. It includes the effects of the following:

- Convection—The effect of source motion relative to the observer, sometimes called the Doppler effect.
- Refraction—The alteration of the sound pressure and directivity as it propagates through a moving fluid with cross-stream gradients.
- Shielding—The decay of the sound as it propagates through portions of the mean-flow gradients where wave-like behavior gives way to exponential decay.

The analytical model does take into account Mach wave radiation associated with eddies convecting at supersonic speeds (which, in turn, tends to amplify the noise levels in the region near the jet axis).

The above analytical model elements have been integrated into a unified computational procedure. The jet plume is subdivided into elemental eddy volumes, each having its own source strength, spectrum, and flow shrouding. The simple closed-form acoustic pressure solutions, combined with the simple aerodynamics calculation method, permit rapid, economical computations of the entire jet plume aerodynamic and acoustic characteristics, including far-field spectra at all observer angles. The contributions from each elemental jet volume, in each frequency band, are simply added based on meansquare pressure. The shock-cell noise contribution is then computed separately and added to the mixing-noise contribution to yield the total far-field spectra.

Noise Suppression Mechanisms

Conventional Bypass versus Inverted-flow Nozzles

Based on the reasonable successes achieved in predicting the aeroacoustic characteristics of a wide variety of nozzle types over a range of operating conditionsit was deemed worthwhile to utilize the theory to analyze the noise suppression mechanisms of coannular nozzles. Of particular interest was how flow inversion (i.e., ducting the high-velocity hot stream to the outside) could provide noise reduction for dual-flow exhaust systems. Theoretical predictions were made of a conventional bypass and an inverted-flow coannular nozzle. The nozzles were sized to give the same thrust and equal primary (high-velocity) and secondary (low-velocity) flow areas. The two nozzles therefore had equivalent thrust, mass flow, primary and secondary stream velocities, and temperatures; thus, the differences in noise should have been solely a function of jet plume profile development and mixing.

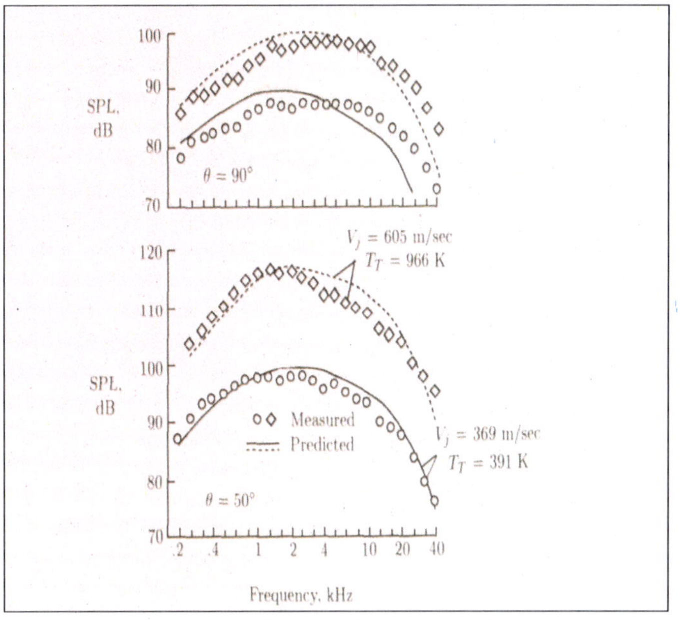

Conical nozzle predicted and measured SPL spectra.

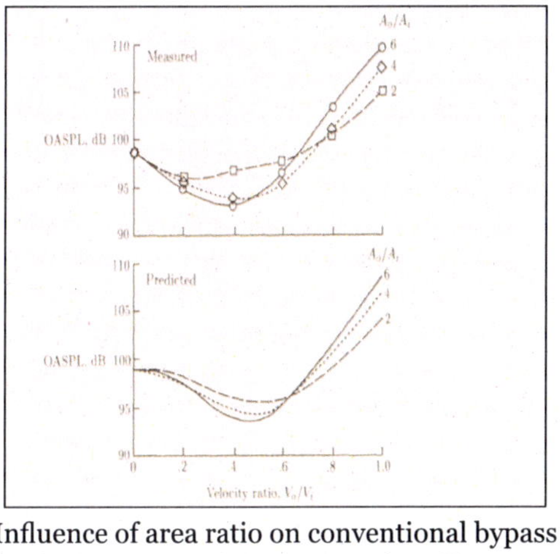

Influence of area ratio on conventional bypass jet OASPL. 9 = 90°; V_i = 305 m/sec; $T_{t,i}$ = 700 K; $T_{T,o}$ = 300 K.

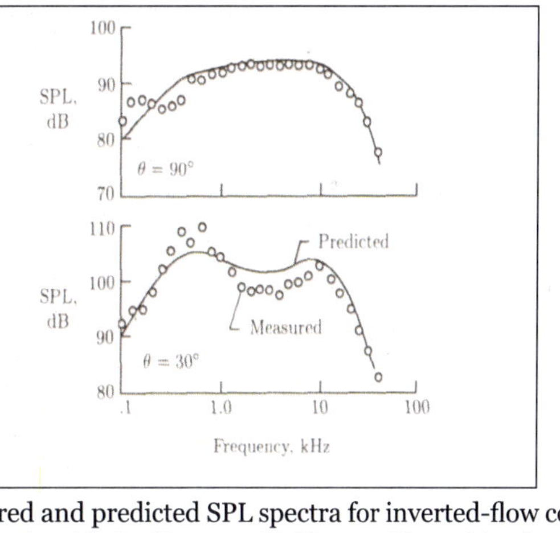

Measured and predicted SPL spectra for inverted-flow coannular nozzle with A_O /A_i = 0.4; V_o/V_i = 1.5; V_i = 366 m/sec; V_o= 549 m/sec; $T_{T,i}$ = 556 K; $T_{T,o}$ = 667 K; R = 12.2 m; D_i = 11.8 cm.

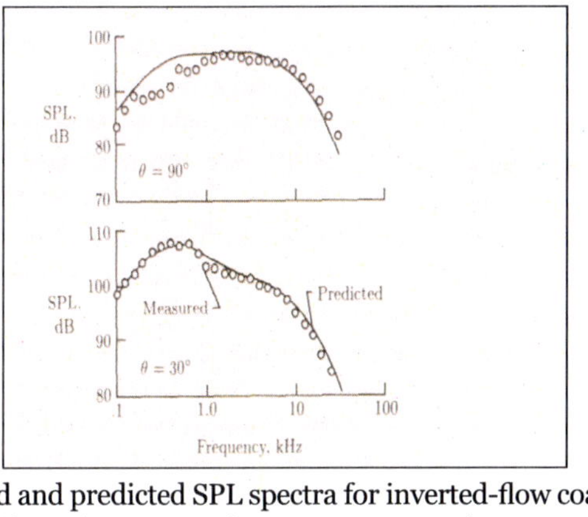

Measured and predicted SPL spectra for inverted-flow coannular nozzle with A_O /A_i = 2.0; V_o/V= 1.5; V_i = 305 m/sec; Vo = 457 m/ sec; $T_{T,i}$ = 294 K; $T_{T,o}$ = 556 K; R = 12.2 m; Di = 11.8 cm.

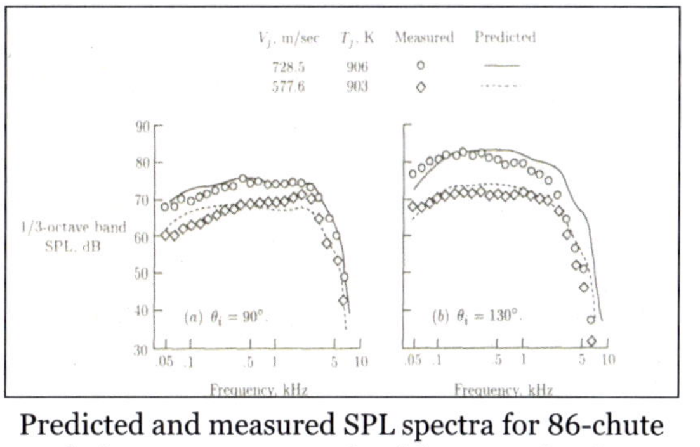

Predicted and measured SPL spectra for 86-chute turbojet suppressor nozzle with area ratio of 2.0.

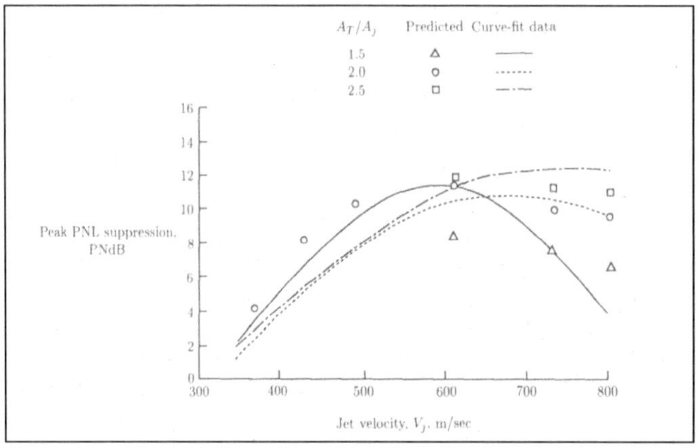

Predicted and measured effect of area ratio on peak PNL suppression for 36-chute turbojet suppressor nozzle.

Aeroacoustic predictions were made for both a conventional bypass and an inverted-flow nozzle for velocity ratio V_s/V_p = 0.7 and area ratio A_s/A_p = 1.0, where subscripts p and s refer to primary and secondary streams. Figure shows mean axial velocity U profiles across the jet plume at several axial stations x/D_p along the plume. The normalized peak velocity at any axial station U_p versus downstream distance is presented in figure for the two nozzle types. These results show that the flow inversion produces more rapid plume decay. Figure shows axial turbulence velocity, a key ingredient in the mixing-noise source strength, at several axial stations along the plume. This figure shows the expected lower turbulence levels at small values of x/D_p for the conventional bypass jet and higher levels of turbulence at large x/D_p than the inverted-flow nozzle.

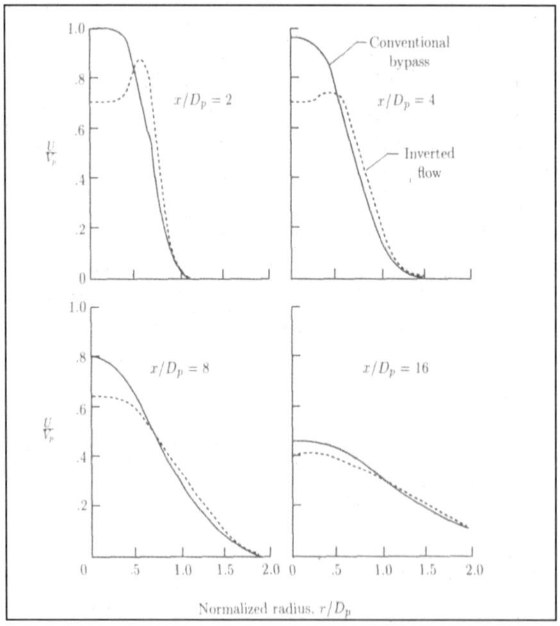

Inverted-flow and conventional bypass mixing mean velocity Profiles.

The corresponding far-field acoustic spectra are shown in figure. At θ = 90° (i.e., in the plane of the nozzle exit), the inverted-flow nozzle exhibits higher noise at high frequencies and lower noise at low frequencies than the conventional bypass nozzle. Since the high-frequency noise generally comes from regions close to the nozzle exit, the highest high-frequency noise correlates with the highest turbulence levels at small values of x/D_p shown in figure. Similarly, low-frequency noise is primarily from the fully developed regions far downstream, and the lowest low frequency noise of the inverted-flow nozzle correlates with its lowest turbulence levels for large x/D_p.

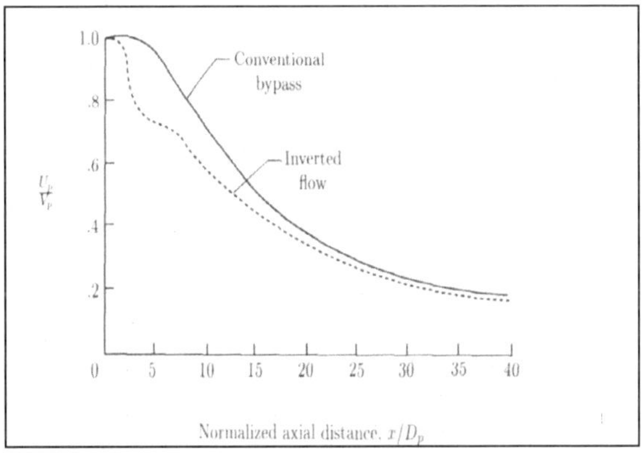

Inverted-flow and conventional bypass mixing peak velocity axial decay.

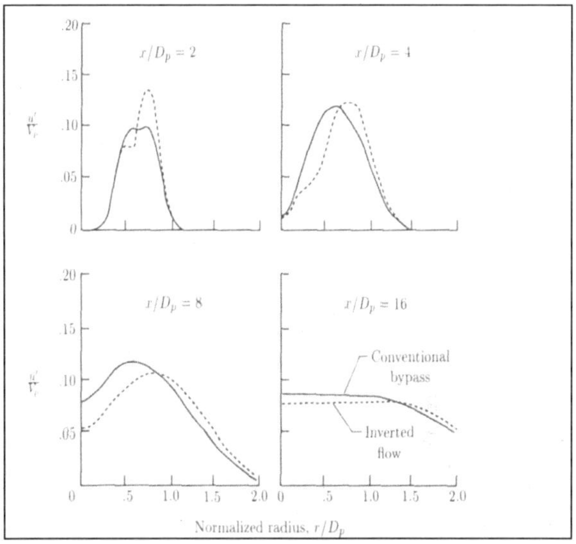

Inverted-flow and conventional bypass mixing turbulence intensity profiles.

The results in figure 12 for θ = 50° (130° from the inlet axis), however, show the inverted-flow jet noise to be lower than that of the conventional bypass nozzle throughout the spectrum. This lower noise is primarily a result of reduced convective amplification, that is, lower eddy convection speeds. The eddy convection speed is proportional to the peak mean axial velocity, and the peak velocity curves shown in figure imply that the inverted-flow jet exhibits lower convection speeds, and therefore reduced convective amplification, than the conventional bypass nozzle. Figure shows nozzle overall sound pressure level (OASPL) versus angle for the two nozzles. Note the shallower slope of the directivity curve for the inverted-flow nozzle, a result of reduced convective amplification.

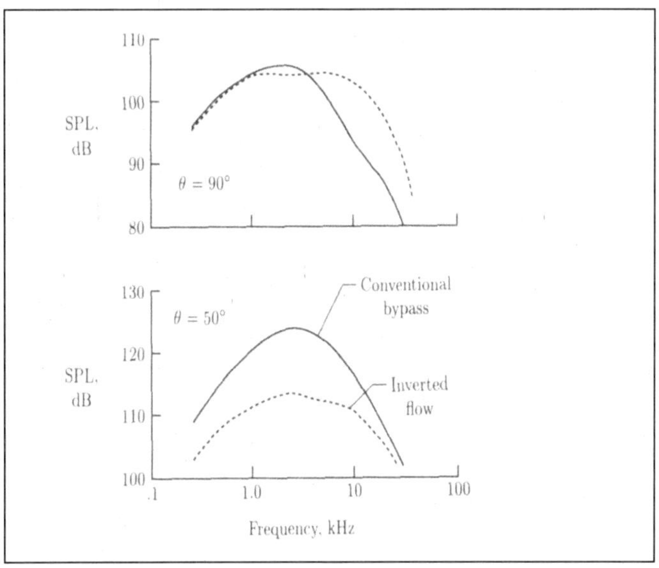

Inverted-flow and conventional bypass mixing SPL spectra.

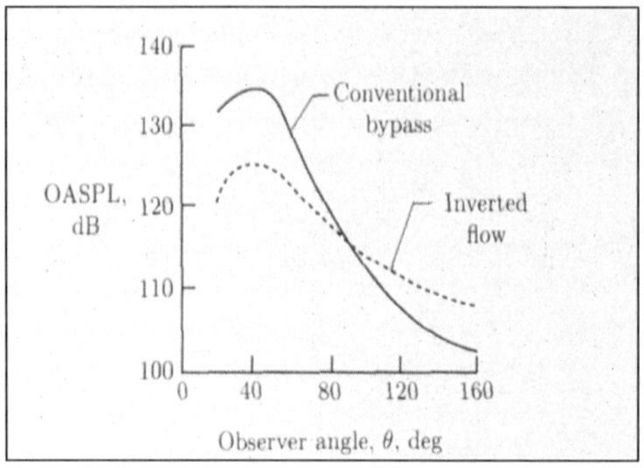

Inverted-flow and conventional bypass mixing OASPL directivity.

To demonstrate the influence of flow shielding as a noise reduction mechanism, the above predictions were repeated with the shielding effects in the acoustic mean-flow interaction model suppressed. These results are shown in figure for θ = 30°. First of all, flow shielding effect is less for the inverted-flow nozzle than for the conventional bypass nozzle, but the reduced convective amplification more than compensates for the loss. The main point to observe is that the shielding effect is substantial at angles close to the jet axis, being on the order of 30 to 40 dB at θ = 30° for these cases. The shielding effect also increases with increasing frequency, as figure shows.

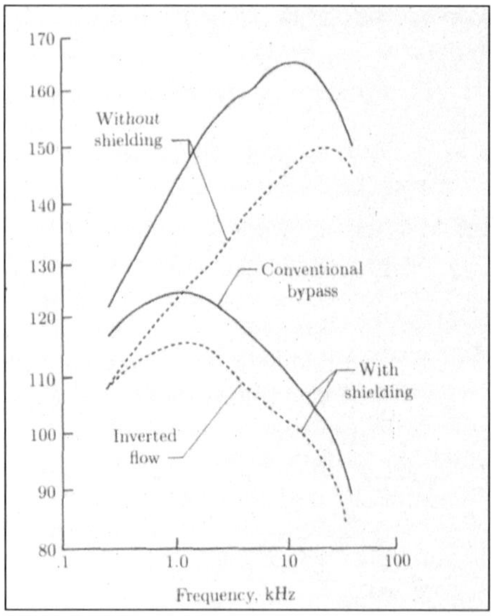

Influence of flow shielding on inverted-flow and conventional bypass nozzle SPL spectrum o = 30°.

Multielement Nozzles

The theoretical concepts for jet noise generation and radiation can be evaluated through use of the analytical model in selective modes. The objective is to identify the relative roles these mechanisms play in the generation and suppression of jet noise. Such a study was carried out with a multielement, 36-chute suppressor nozzle used as a representative case study. The basic jet noise mechanisms are as follows:

- Turbulent-mixing-noise generation.
- Convective amplification.

- Fluid shielding.
- Shock-cell noise.

It is of interest to evaluate how the individual mechanisms combine to yield the far-field result and, more importantly, how these mechanisms change due to the addition of a suppressor to a baseline nozzle.

A parametric study was performed to evaluate the relative contributions of the four mechanisms to the far-field noise for both a baseline conical nozzle and a typical high-suppression, multielement nozzle. A 36-chute turbojet suppressor with a ratio of total nozzle area A_T to flow area A_j of 2.0 was chosen for this study as representative of a high-element-number, high-suppression (10 to 12 PNdB) exhaust system. A typical take off condition of V_j = 732 m/sec (2400 ft/sec) and T_j = 906 K (1630°R) was selected for evaluation. The exit area of both the baseline conical nozzle and the 36-chute suppressor nozzle was 0.218 m² (338 in²). Noise characteristics were predicted for a 732-m (2400-ft) sideline distance.

The prediction computations were performed in the following four modes: (1) complete acoustic calculation; (2) as in (1), but with shock-cell noise omitted; (3) as in (2), but with fluid shielding omitted; and (4) as in (3), but with convective amplification omitted. (For the 36-chute suppressor, mode (1) was omitted since the ability to model shock-cell noise of multichute nozzles was not established at the time these computations were performed. However, Stone has evolved a semiempirical prediction method for shock-cell noise of multielement suppressor nozzles utilizing more recent data.The difference in noise levels between modes (1) and (2) is a measure of the shock-cell noise contribution to the total jet noise signature. The difference in noise levels between modes (2) and (3) is a measure of the influence of fluid shielding on the jet noise. Finally, the difference in noise levels between modes (3) and (4) indicates the amount of convective amplification that is present in the jet.

The results of the above series of computations are summarized in figures. Figure shows the perceived noise level (PNL) directivity patterns for the different prediction modes, which should be compared with mode (1) predictions (mode (2) for the 36-chute nozzle). Figures show the corresponding spectrum shapes (1/3-octave SPL) at 50° and 130° relative to the inlet axis.

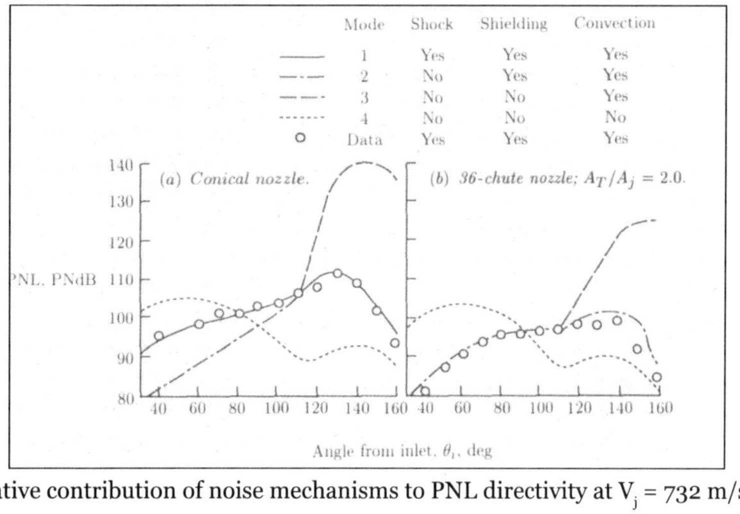

Relative contribution of noise mechanisms to PNL directivity at V_j = 732 m/sec.

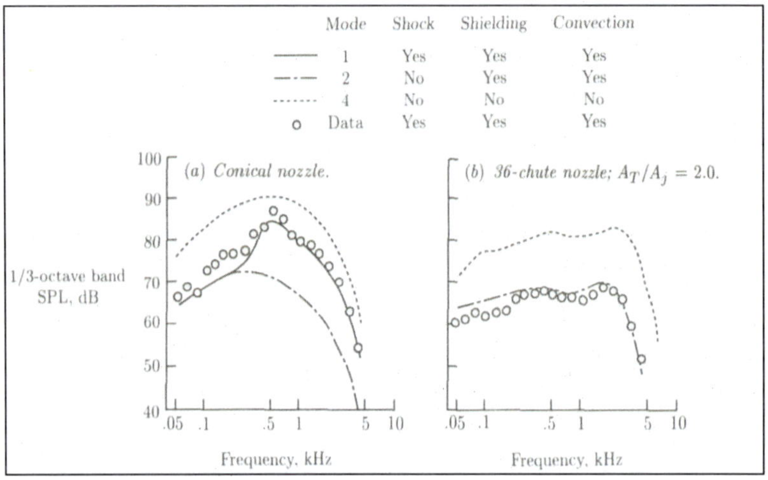

Relative contribution of noise mechanisms to SPL spectra at $\theta_i = 50°$ for conical nozzle and 36-chute nozzle at V_j = 732 m/sec.

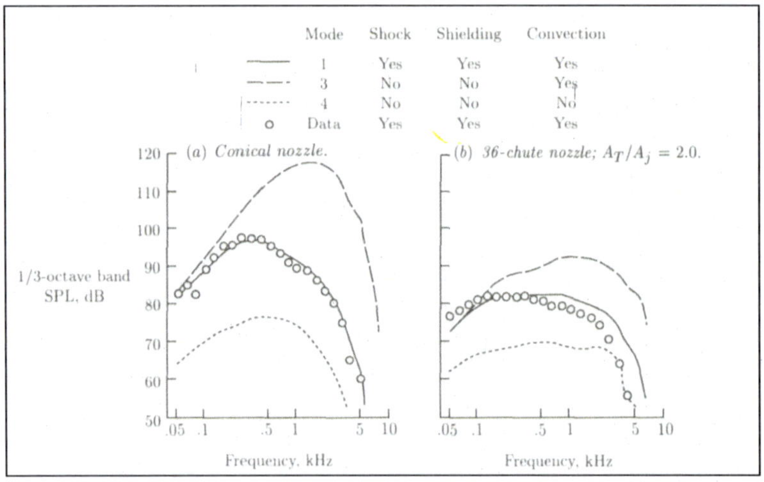

Relative contribution of noise mechanisms to SPL spectra at Bi = 1300 for conical nozzle and 36-chute nozzle at V_j = 732 m/sec.

For the conical nozzle PNL directivity patterns, the shock-cell noise contributes substantially to the total noise in the forward quadrant, that is, at angles to the inlet less than 90°. This contribution can be seen in the difference between mode (1) and (2) predictions. There is no contribution of shock-cell noise close to the jet axis (i.e., at angles greater than 120° from the inlet axis), since predictions for modes (1) and (2) are the same in this region. There is no fluid shielding for observer angles less than about 110°, based on a comparison of predictions for modes (2) and (3). For angles greater than 110°, however, shielding effects become quite substantial, on the order of 30 PNdB. Eddy-convection effects are also large; they increase the noise in the aft quadrant (angles greater than 90°). This effect is apparent from comparing predictions for modes (3) and (4).

The prediction for mode (4) shown in figure represents the basic turbulent mixing noise in the absence of convection and fluid-shielding effects. It possesses a basic, non-constant directivity pattern dictated by the weighted summation of various quadrupole types composing the turbulent eddies. This basic pattern is only symmetric about 90° when the local flow Mach number is zero because the quadrupole weighting factors are a function of local Mach number and bias the radiation toward the forward quadrant.

The corresponding PNL directivity patterns for the 36-chute suppressor are shown in figure. The trends discussed above for the conical nozzle are qualitatively similar for the 36-chute nozzle, with the exception of the shock-cell noise contribution. The predictions were made with the shock-cell noise neglected (mode (2)), and yet the predictions agree well with the data, as figure shows. This finding implies that shock-cell noise is not a significant feature of a high-element-number multichute nozzle. It also appears that neither convection effects nor fluid-shielding effects are as strong as for the conical nozzle.

The breakdown of mechanisms for a typical forward-quadrant angle of 50° is shown in figure. No shielding occurs at this angle; therefore, the mode (3) results are omitted, as they are identical to the mode (2) results. The conical-nozzle results show an interesting counteraction among the mechanisms. The basic mixing-noise spectrum (mode (4)) yields a high noise level, much higher than the measured level. The convection effect is to Doppler shift and drop this spectrum to a level significantly lower than the data (except at very low frequencies), as indicated by the mode (2) prediction. Finally, the addition of the shock-cell noise spectrum raises the spectrum back to the measured level at middle to high frequencies.

The corresponding spectrum results for the 36-chute nozzle are shown in figure. The good agreement between the spectrum prediction and the measured spectrum for mode (2) substantiates the implication from figure: shock-cell noise may not be a significant source for a high-element-number multichute suppressor. However, acoustic data for multielement suppressors employing fewer elements (e.g., 20) show evidence of shock-cell noise in the forward quadrant, particularly in flight. Again, the effect of convection is to reduce the level and Doppler shift the spectrum to lower frequencies.

Near the peak noise angle of 130°, convection effects are significant. They produce a dramatic amplification of the mixing noise, as the results in figure show. Another counteraction of mechanisms occurs at this angle and involves the competing effects of convection and fluid shielding. The basic mixing-noise spectrum (mode (4)) is much lower than the measured level, as shown in figure. The effect of convection is to increase the levels by as much as 40 to 50 dB at high frequencies. The effect of shielding, however, is to reduce the noise levels by 20 to 30 dB at high frequencies, consistent with the measured levels.

It is interesting to note that the convection effect Doppler shifts the basic mixing noise spectrum to higher frequencies, as would be expected from classic notations of moving-source acoustics. However, the fluid-shielding effects, which increase with increasing frequency, attenuate the high-frequency portion of the convected spectrum to such a large extent that the resulting spectrum peaks at a much lower frequency, lower than even the basic unconverted spectrum peaks. This attenuation explains the observed "reverse Doppler shift" at angles close to the jet axis.

The competing influences of convection and fluid shielding are also evident in the predictions for the 36-chute nozzle shown in figure. The magnitudes of these effects are considerably smaller than those exhibited by the conical nozzle. For example, at 2000 Hz the convective amplification is 22 dB for the 36-chute nozzle, compared with 48 dB for the conical nozzle at the same frequency. Similarly, the fluid-shielding attenuation is only 12 dB at 2000 Hz for the chute nozzle, compared with 31 dB of attenuation for the conical nozzle.

The various suppression mechanisms can be isolated explicitly by examining the difference between the various prediction curves shown in figures, the total PNL suppression as a function of angle is the difference between the conicaland 36-chute-nozzle total noise PNL directivity patterns. This difference can be compared with measured PNL suppression and is shown in figure. The predicted total PNL suppression compares well with the measured suppression.

For the results of figures, it can be concluded that the high-element number multichute nozzle almost completely suppresses the shock-cell noise and thus permits the mixing noise to dominate the forward-quadrant spectra. The shock-cell noise suppression is then approximated by the difference between mode (1) and mode (2) conical-nozzle predictions. This estimated shock-cell noise suppression is higher than the total predicted suppression in the forward quadrant; hence, some other mechanism is providing negative suppression that is, increasing the PNL.

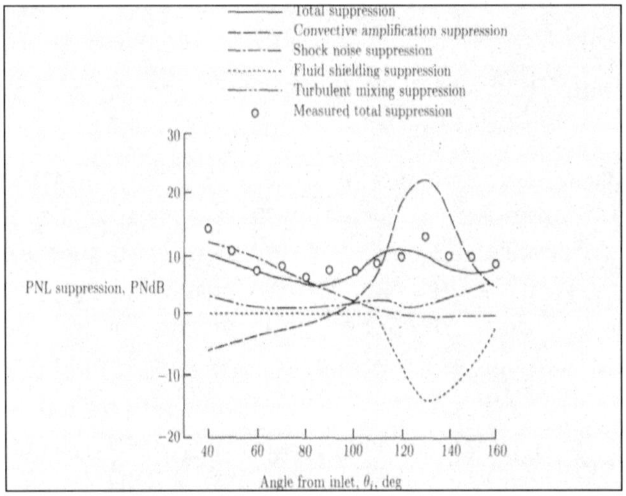

PNL suppression composition for 86-chute nozzle at V_i = 732 m/sec. A_T/A_j = 2.0.

The suppression due to convective amplification can be computed by first calculating the convective amplification for each nozzle (mode (3) PNL minus mode (4)) and then subtracting the 36-chute-nozzle result from the conical-nozzle result. The convective amplification suppression is shown in figure. Note that it is negative in the forward quadrant; this explains why the shock-cell noise suppression is greater than the total (net) suppression.

Similarly, the difference between the fluid-shielding attenuation for the 36-chute suppressor and the conical nozzle was computed from the results shown in figure, and this difference is also shown in figure. From this result, it is apparent that a multielement suppressor exhibits less fluid-shielding effects than a conical nozzle, that is, part of the beneficial effect of fluid shielding is lost by the addition of a suppressor.

The observed suppression in the aft quadrant is primarily a result of reduced convective amplification, offset somewhat by a loss in fluid shielding. This delicate balance between convection and shielding effects in the aft quadrant is very difficult to predict accurately because these two effects are large in magnitude but opposite in sign, as illustrated in figure.

Finally, the suppression of basic turbulent-mixing-noise generation has been evaluated by subtracting the prediction for mode (4) for the 36-chute nozzle from that for the conical nozzle, and this result is also shown in figure. The basic mixingnoise suppression is quite small, from 1 to 5 dB over the range of angles shown, and this result is contrary to historical conceptions of how much multielement suppressors suppress jet noise.

The multichute suppressor in fact generates approximately the same total mixing noise as the equivalent conical nozzle but redistributes the noise to higher frequencies. This is dramatically illustrated in figure, in which the basic mixing-noise spectra (mode (4)) for the two nozzles are presented. Also shown are these same spectra with the atmospheric air attenuation removed (i.e., the "lossless" spectra). The multichute lossless spectrum has about the same peak level as the conical-nozzle lossless spectrum, but at a much higher frequency. The ratio of 36-chute-nozzle peak-noise (lossless) frequency to conical-nozzle peak-noise (lossless) frequency is about 6:1. This is precisely the ratio of the conical-nozzle diameter to 36-chutenozzle equivalent-area diameter.

The reduction in shock-cell noise produced by a multichute suppressor can be explained by the fact that breaking up a large, round jet into very small, discrete, rectangular jets will cause the shock-cell formation to be dissipated much more rapidly. The shock-cell spacings and cross-sectional dimensions will be much smaller, and the cells are likely to be less numerous. The resulting radiation is therefore likely to be much lower in level and higher in frequency than that for a conical nozzle.

The observation that the total generated mixing noise is not significantly different for a multichute suppressor than for a conical nozzle is explained by the fact that the multichute-nozzle mixing layer perimeter close to the nozzle exit plane is considerably larger than that of an equivalent-area conical nozzle. The high frequency noise generated in the initial shear layers should therefore be higher by the ratio of perimeters, provided the pre-merged portions of the chute mixing layers have approximately the same turbulence characteristics. Once the chutes have merged, a large, axisymmetric jet forms which has a substantially lower velocity than the exit value; therefore, the multichute-nozzle low-frequency noise levels should be lower than the corresponding conical-nozzle levels.

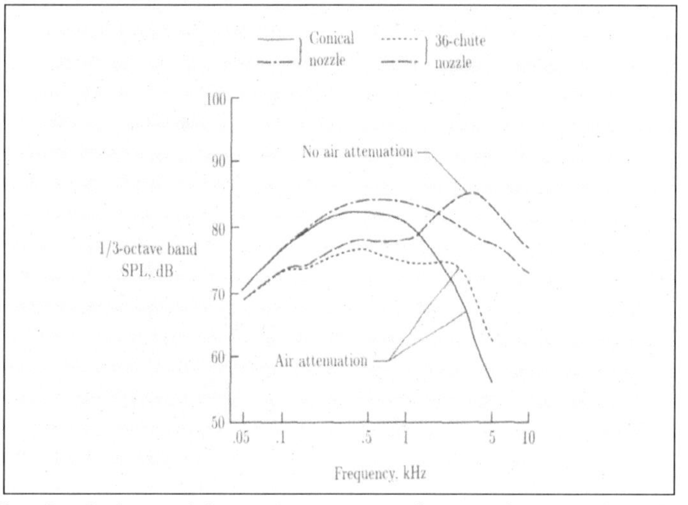

Predicted turbulent-mixing-noise spectra at $\theta_i = 90°$ for conical nozzle and 36-chute nozzle ($A_T/A_j = 2.0$) at $V_j = 732$ m/sec.

The reduction in convection effects exhibited by a multichute nozzle is the result of lower eddy-convection velocities. The rapid decay of plume mean velocity exhibited by a multichute nozzle suggests that the majority of the noise-producing turbulent eddies in the plume are convecting downstream at a substantially lower velocity than in a conical nozzle.

The reduced fluid-shielding effects characteristic of a multichute nozzle can also be related to the rapid plume velocity (and temperature) decay. Fluid shielding increases with increasing plume flow velocity and temperature; therefore, the lower velocity and temperature levels resulting from the rapid chute-element mixing provide less fluid shielding than those of an equivalent-area conical nozzle.

From the preceding considerations, it can be concluded that the best suppression is achieved by producing the most rapid plume decay. Higher numbers of elements and higher area ratios give more rapid decay. Area ratio seems to control the velocity level plateau formed by the merging of the individual chute element jets into a single annular jet. Higher area ratios yield lower merged-velocity levels and hence lower convection speeds. Element number seems to control how fast the merged-velocity level is reached; higher numbers of elements give more rapid decay of the individual element jet flows to the merged-velocity level. One additional benefit of higher numbers of elements is that the initial mixing-noise generation is pushed to higher frequencies by virtue of the smaller turbulence scales associated with the smaller, individual jet elements.

Jet Noise Suppression Concepts

Various experimental studies have demonstrated the noise reduction potential of jet noise suppression concepts which essentially alter the jet plume development. All these concepts may be grouped under the following two broad categories:

- Geometric concepts, such as multielement suppressors (e.g., chutes, spokes, and tubes) at the nozzle exit plane, ejectors surrounding the jet in the vicinity of the nozzle exit, annular plug or 2-D nozzles, and combinations of the above.
- Aerothermodynamic concepts, such as inverted-velocity profile (i.e., a higher velocity jet surrounding a lower velocity jet), thermal acoustic shield (i.e., surrounding either fully or partially the main jet with a jet having a high static temperature and low velocity), and shock noise control by appropriate contouring of the nozzle flow path.

Some jet noise suppression concepts have employed a combination of both geometric and aerothermodynamic concepts (e.g., a dual-flow, multielement suppressor with inverted-velocity profile).

Geometric Concepts

Multielement Suppressors

The principal jet noise suppression mechanism underlying the multielement suppressors (such as chutes, spokes, or tubes) is the division of the large single jet into many small jets with sufficient separation between these jets to enhance mixing with the ambient air to yield a rapid decay of the jet plume. Mixing is enhanced by the increased surface area of the jet available for shear with the ambient air, a classic momentum diffusion concept. Also, since each small jet has a much smaller equivalent diameter than a circular jet, its effectiveness as a noise radiator is restricted to a much smaller physical length compared with the single large jet, as long as the separate jets do not merge. The merged jet will have a much lower jet velocity. This lower velocity results in the acoustic energy being transferred to smaller scale turbulent eddies which, by virtue of their small size, generate noise in the higher frequency ranges than the turbulent eddies of a large single jet. This higher frequency noise generation results in a suppressor spectrum which has high and low frequency peaks, unlike the conical nozzle, which has a low-frequency peak only. The acoustic signal suffers atmospheric absorption in reaching the far-field observer. As various data and theories have indicated, high-frequency acoustic waves suffer significantly more atmospheric attenuation than low-frequency waves over the same distance. Since multielement suppressors have a larger contribution of acoustic energy in the high-frequency region and that energy suffers much higher levels of atmospheric absorption, the far-field noise of these suppressors is lower than that of single conical nozzles.

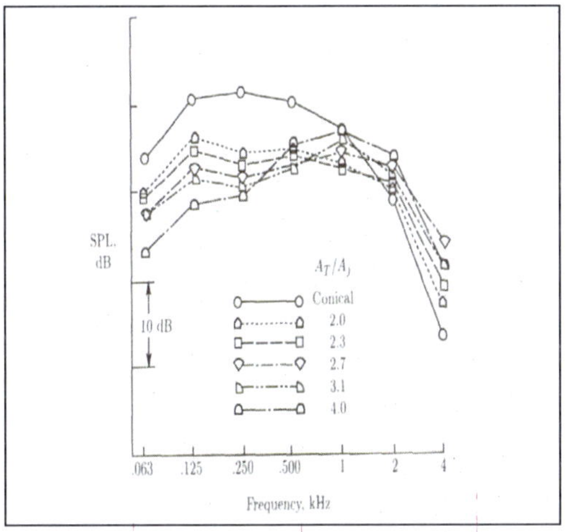

Spectra illustrating dual-peals characteristics of multi-tube nozzles. NPR≈2.1; $T_T \approx 1520°R$; $V_j \approx 1870$ ft/sec.

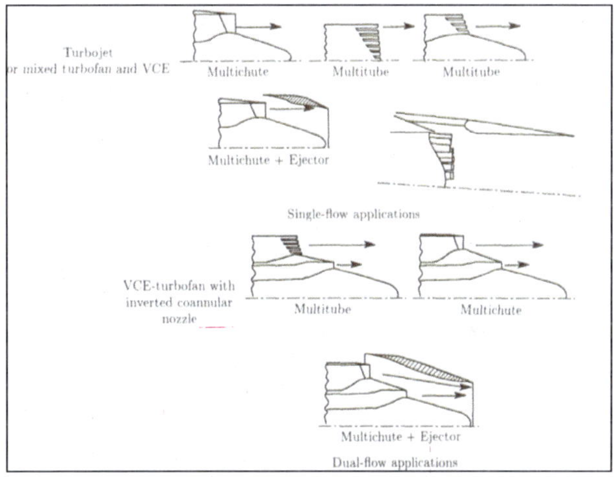

Examples of multielement suppressors with and without ejectors.

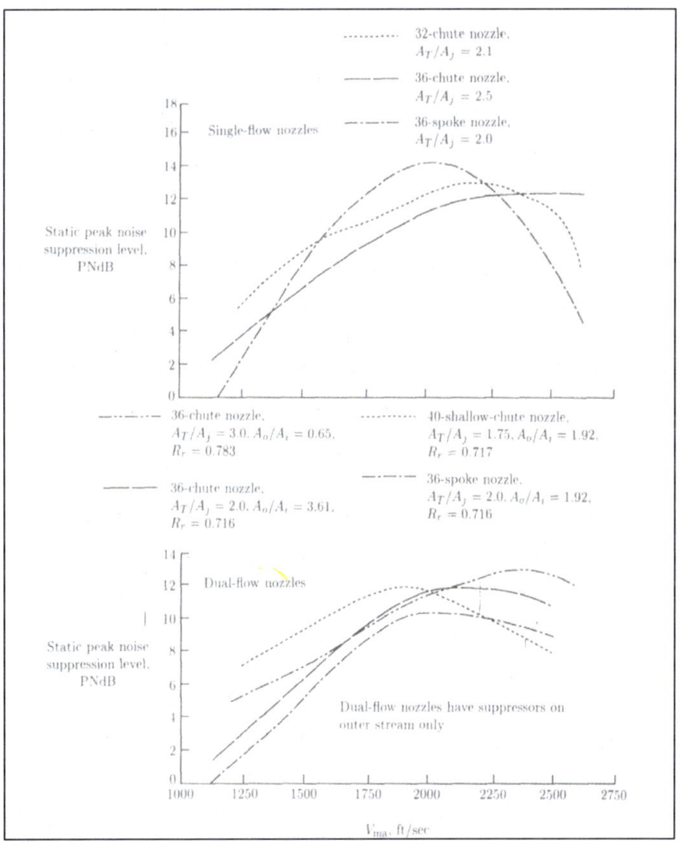

Typical static peak/PNL suppression characteristics of multielement suppressors. 2400-ft sideline; standard day (59°, 70 percent relative humidity); AT = 338 in².

Figure shows the following two families of multielement suppressors, for which a large body of data exists:

- Single-flow applications for a turbojet or mixed turbofan and variable cycle engine (VCE).
- Dual-flow applications with a suppressor in the high-velocity stream for a turbofan and VCE with inverted-velocity profile.

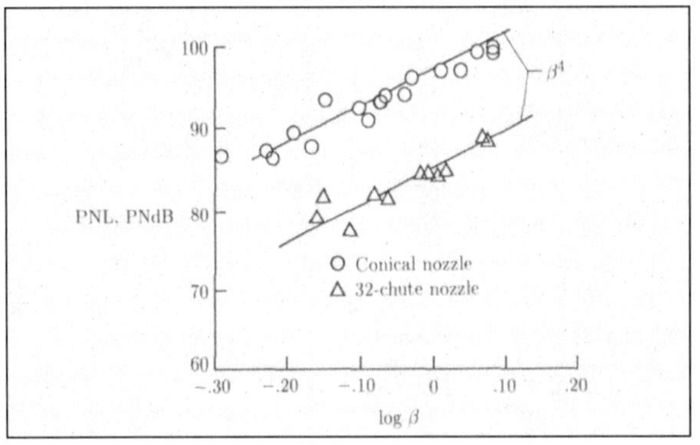

PNL at θi = 50°.

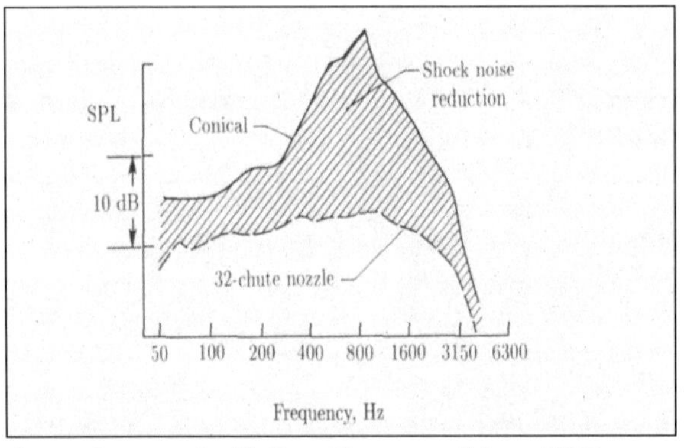

(b) SPL at θ_i = 50°; 2400-ft sideline; V_j = 2300 ft/sec; NPR = 3.2; J-79 size.
Forward-quadrant shock noise reduction of 32-chute suppressor and conical nozzle.

Figure summarizes the suppression potential of various single- and dual-flow applications of multielement suppressor concepts in terms of static peak PNL suppression measured over the range of mass-averaged jet velocities V_{ma} covering typical take off, thrust cutback, and approach conditions. Single-flow applications generally exhibit slightly higher suppression than outer-stream suppressors on dual flow systems, as the complete jet is segmented for single-flow applications. As an illustration of the potential of multielement suppressors for shock noise reduction, figure shows the forward-quadrant noise reduction in terms of PNL at Bi = 50° for a wide range of supersonic jet Mach numbers. Also shown are the spectral reductions for a 32-chute, single-stream suppressor compared with those for a conical nozzle.

Next, data trends obtained with geometric variations of multitube, multispoke, and multichute suppressors are presented. For multitube suppressors, the design variables having first order impact are:

- Suppressor area ratio (total area/flow area, AT/Aj).
- Number of tubes.
- Ratio of internal tube length to diameter, Lt/Dt.
- Exit plane and base plane stagger.

The influence of the above four design variables on both acoustic and aerodynamic performance (in terms of pressure levels in the base region) is shown in figures. The following trends are indicated.

At high jet velocities, higher area ratios yield higher levels of acoustic suppression than lower area ratios yield; at low jet velocities, the inverse is true.

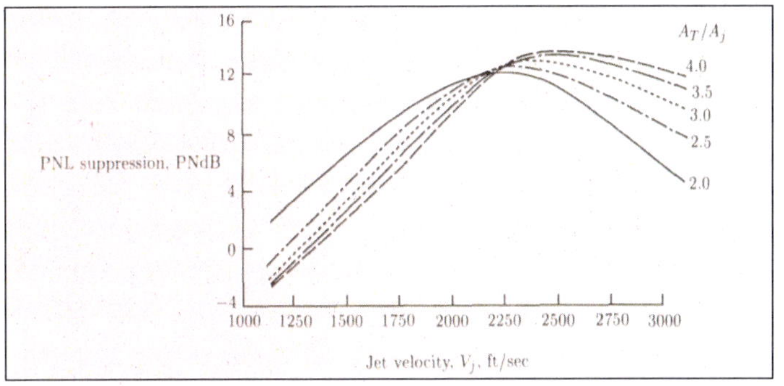

PNL suppression; 1500-ft sideline; no shroud; based on data at A_T/A_j= 2.0, 2.3, 2.7, 3.1, and 4.0 (2.0 and 4.0 repeated).

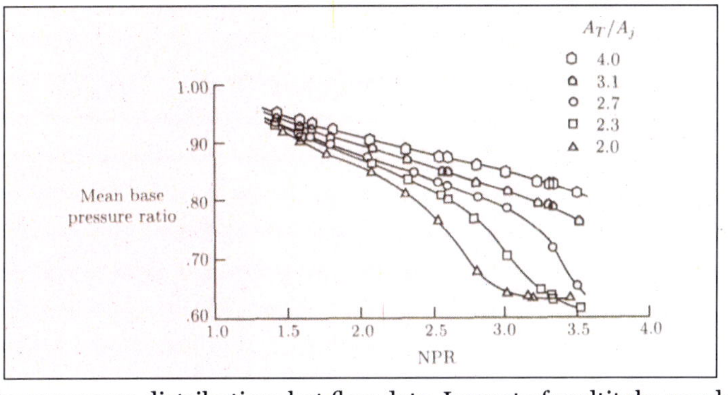

Base pressure distribution; hot flow data. Impact of multitube nozzle area ratio on acoustic and aerodynamic performance. L_t/D_t = 2.2.

The designs with higher area ratios enhance ventilation of the base region to ambient air, the result being base pressures which approach the ambient pressure and thus base drag is reduced. However, increasing the area ratio requires a larger nozzle diameter and associated weight and drag penalties.

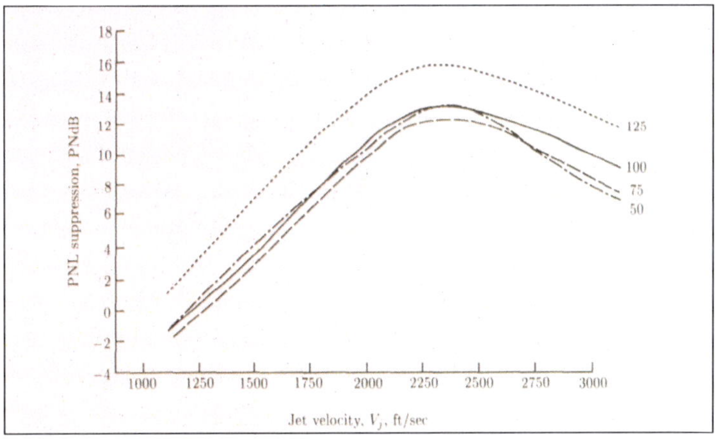

Effect of number of holes; 1500-ft sideline; no shroud; based on data for 55, 85, and 121 holes (55 holes repeated).

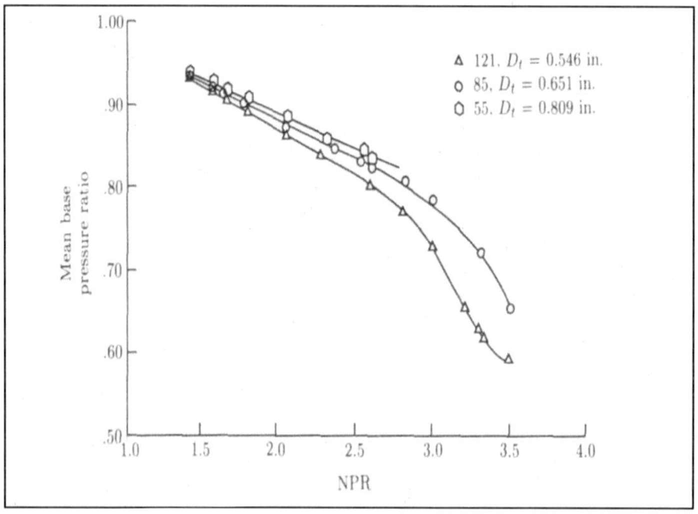

Impact of multitube nozzle degree of segmentation on acoustic and aerodynamic performance. $A_T/A_j = 2.7$; $L_t/D_t = 2.2$.

For a given area ratio, increasing the number of tubes results in higher levels of noise suppression because of the shift of acoustic energy into higher frequencies. However, the increase yields lower base pressure for tubes in the interior because of poor ventilation and results in high base drag.

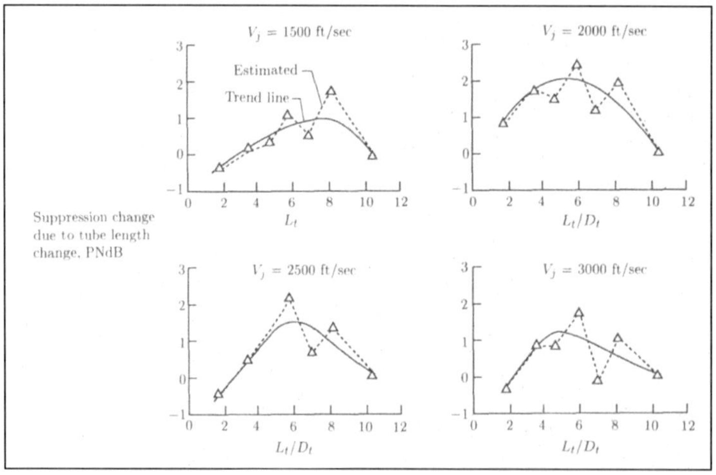

Effect of tube internal length ratio on noise suppression; 300-ft sideline.

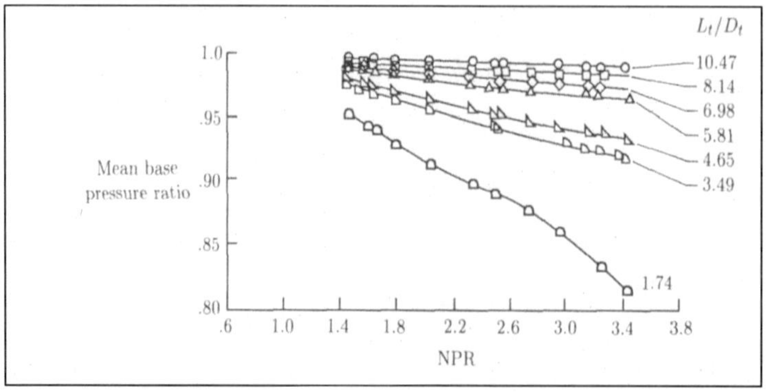

Base pressure distribution. Impact of multitube nozzle ratio of tube length to diameter on acoustic and aerodynamic performance. $A_T/A_j = 3.19$; 85 tubes; no shroud.

The data for the ratio of tube length to diameter indicate optimum PNL suppression as a function of jet velocity. Long tubes enhance base ventilation by virtue of the distance available for entrainment of ambient air. However, weight and stowing for cruise determine the upper limit for this parameter.

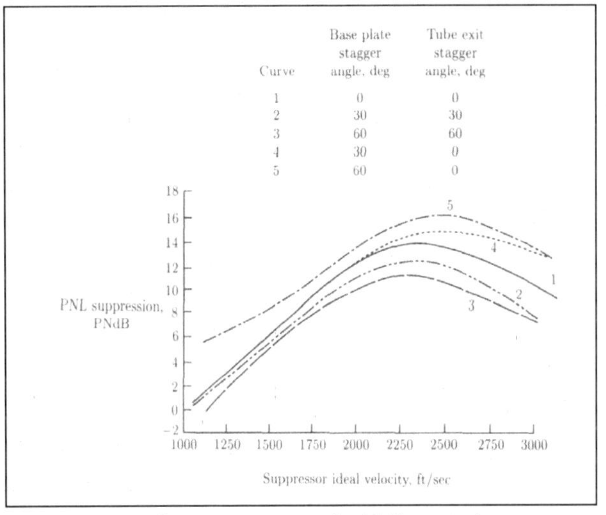

Acoustic performance; 1500-ft sideline; $A_T/A_i = 2.27$.

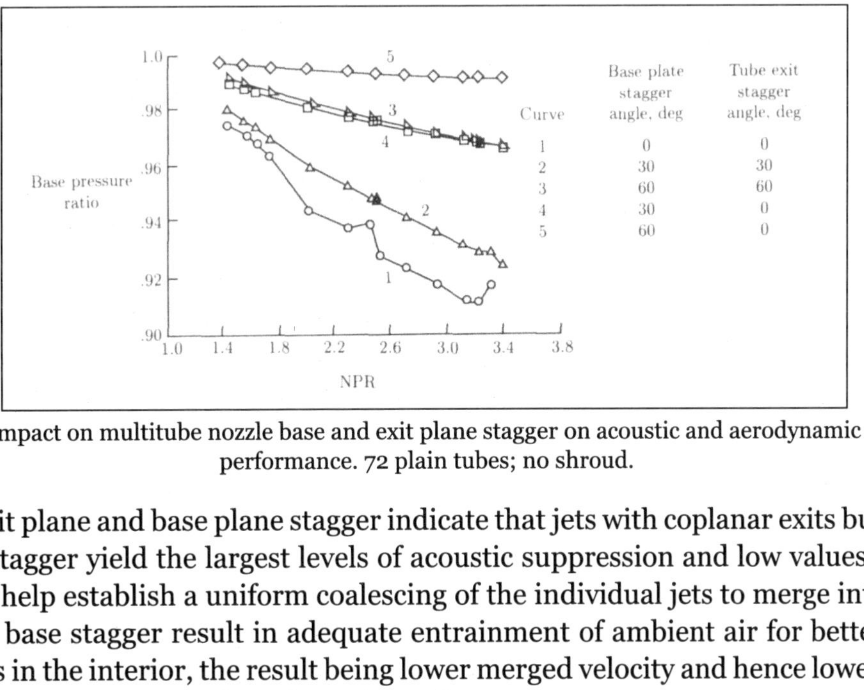

Impact on multitube nozzle base and exit plane stagger on acoustic and aerodynamic performance. 72 plain tubes; no shroud.

The data for exit plane and base plane stagger indicate that jets with coplanar exits but large values of base plane stagger yield the largest levels of acoustic suppression and low values of base drag. Coplanar exits help establish a uniform coalescing of the individual jets to merge into a single jet. High values of base stagger result in adequate entrainment of ambient air for better mixing and diffusion of jets in the interior, the result being lower merged velocity and hence lower noise levels. Improved entrainment also results in lower base drag.

Chutes and spokes have similar geometric planforms, the only difference being that chutes have the capability of entraining the ambient air by means of the ventilation channel upstream of the nozzle exit plane and spokes do not. Hence, the multispoke suppressors yield lower aerodynamic performance than the multichute suppressors. Acoustic suppression potential of chutes and spokes is very similar. Figure is a brief summary of the multichute and multispoke nozzle acoustic and aerodynamic performance trends with key geometric parameters, such as suppressor area ratio A_T/A_p element number, and shape.

Chutes exhibit similar noise suppression trends with suppressor area ratio as the tubes, that is, higher area ratios yield more suppression at high jet velocities and the lower area ratios yield more suppression at lower jet velocities. However, as the suppressor area ratio is increased, aerodynamic performance of chutes in terms of gross thrust coefficient' $C_{f,g}$ deteriorates, a trend opposite to that of tubes. The principal reason for this deterioration is that as area ratio is increased for chutes, the base area is increasing for a given area available for entrainment of ambient air through the entrainment channels. Acoustic suppression is a fairly weak function of number of chutes. For a given suppressor area ratio and chute depth/width ratio D/W, the aerodynamic performance of the suppressor decreases with increases in chute number. The principal reason for this decrease is that as the chute number increases for a given suppressor area ratio, the chute widths decrease. Thus for a constant chute depth/width ratio, the chute depth decreases, the result being poorer ventilation of the chute in the interior and hence an increase in base drag. Acoustic suppression levels obtained with chutes and with spokes are similar, with a slightly higher level of suppression exhibited for chutes at high jet velocities.

Exit plane angle (canting) has a noticeable impact on both acoustic and aerodynamic performance. A 10° cant helps the individual jets to retain their identity for a longer distance and thus helps improve shearing of the multiple jets by the ambient air to yield higher levels of acoustic suppression. However, a 10 o cant tends to separate the flow from the plug surface, the result being poor aerodynamic performance.

Planform shape of the spokes (i.e., tapered or parallel-sided) affects both acoustic and aerodynamic performance. Parallel-sided spokes yield a larger flow perimeter for shearing by ambient air and thereby yield higher levels of acoustic suppression than the tapered spokes. Aerodynamic performance of parallel-sided spokes is lower than that of tapered spokes since parallel-sided spokes have larger base areas in the interior, where ambient air cannot conceivably penetrate.

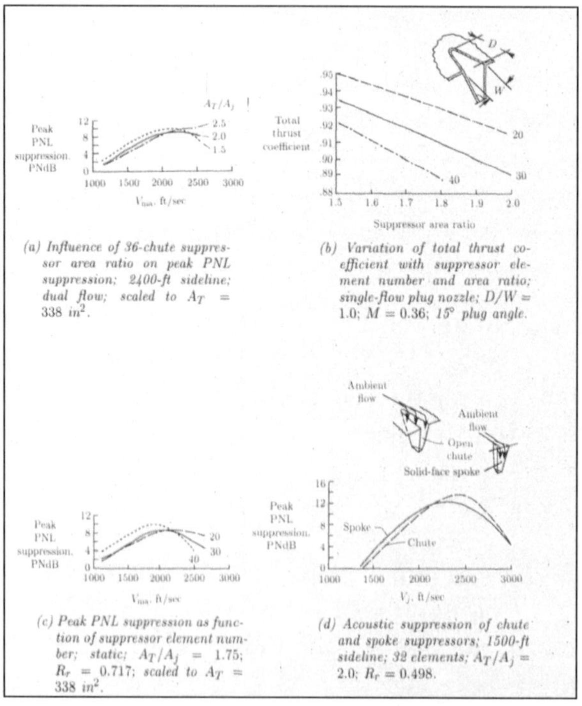

(a) Influence of 36-chute suppressor area ratio on peak PNL suppression; 2400-ft sideline; dual flow; scaled to A_T = 338 in^2.

(b) Variation of total thrust coefficient with suppressor element number and area ratio; single-flow plug nozzle; D/W = 1.0; M = 0.36; 15° plug angle.

(c) Peak PNL suppression as function of suppressor element number; static; A_T/A_j = 1.75; R_r = 0.717; scaled to A_T = 338 in^2.

(d) Acoustic suppression of chute and spoke suppressors; 1500-ft sideline; 32 elements; A_T/A_j = 2.0; R_r = 0.498.

Figue: Multichute and multispoke suppressor nozzle acoustic and aerodynamic performance trends.

Figure summarizes the acoustic suppression versus thrust degradation for various tube, spoke, chute, plug, coannular, and ejector nozzles. Cruise performance considerations certainly will require that the suppressors be stowed to minimize mission impact. Typical schemes for stowage and deployment of chutes and tubes are conceptually illustrated in figure.

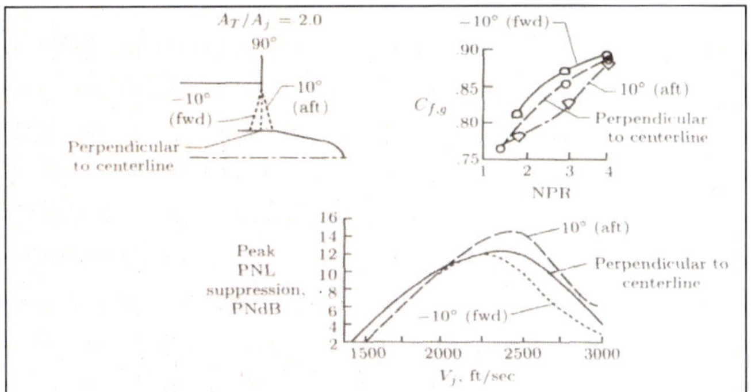

(e) Influence of element cant angle on acoustic and aerodynamic performance.

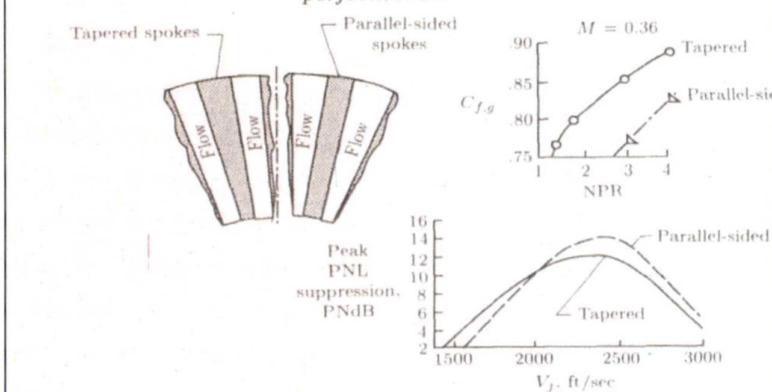

(f) Influence of element planform shape on acoustic and aerodynamic performance.

Concluded.

The above discussion indicates that multielement suppressor design involves a complex interaction of acoustic suppression potential, aerodynamic performance, and mechanical feasibility (i.e., implementation, stowability, and reliability). The ultimate impact of these considerations on the mission is equally important.

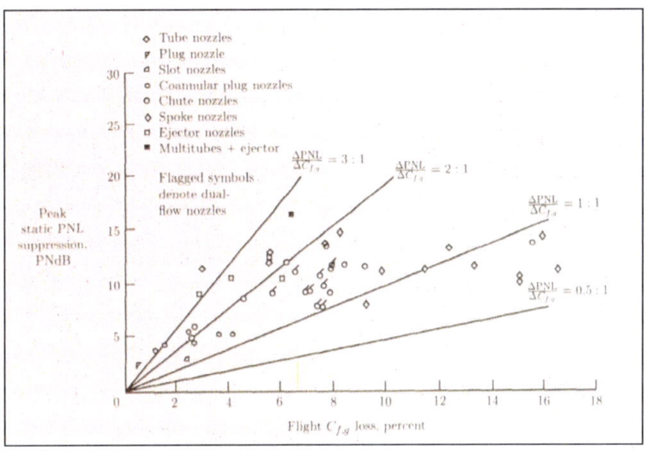

Generalized flight performance and static suppression levels relative to conical nozzle. Based on measured aerodynamic performance and acoustic data. Nominal conditions: single flow—V_{ma} ≈ 2400 ft/sec and NPR ≈ 3.0; dual flow—V_{ma} 2200 ft/sec, NPR_o ≈ 3.0, and NPR_i ≈2.5 → 3.5.

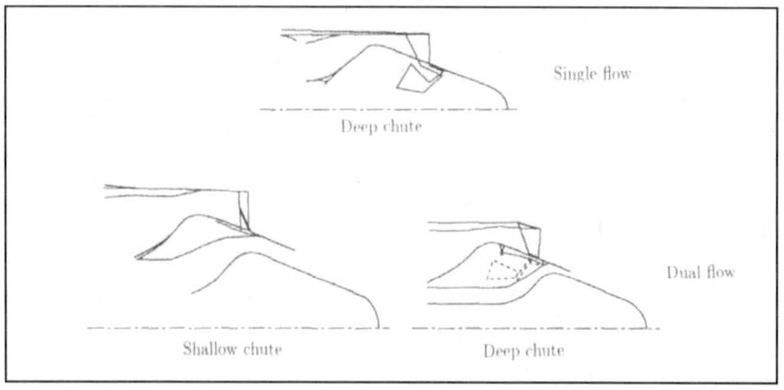

Installation schemes for stowed and deployed multielement suppressors.

Ejectors

An ejector, when designed as an integral part of the propulsive nozzle system, can yield optimum aerodynamic performance for the entire mission by providing an optimum flow expansion surface. It can also provide additional noise suppression than a nozzle system without an ejector by virtue of its capability for induction of ambient air to enhance mixing and shielding or suppression of noise sources within the ejector length. For an extensive discussion of the principles of noise reduction by mixing nozzles and ejector nozzles, see reference. Ejectors used in conjunction with conical nozzles or basic coannular plug nozzles have been ineffective since the dominant acoustic sources are located outside the ejector. However, for suppressor nozzles, the dominant acoustic sources are located close to the jet exit plane; hence, ejectors with acoustic treatment on the internal flow surface effectively improve the basic suppressor acoustic performance. Air induction by ejectors is also more effective, as induced air can impact mixing in the region of dominant acoustic sources.

Flight tests on a Viper turbojet engine fitted with various exhaust suppressors with and without treated ejectors are reported in reference 23. A maximum attenuation of 14 EPNdB was measured at an ideal jet exhaust velocity of 732 m/sec (2400 ft/sec) at an adjusted altitude of 152 m (500 ft).

Influence on ejector spacing S from the suppressor exit plane (normalized by the equivalent diameter based on flow area D,q) on peak PNL as a function of mass-averaged jet velocity is shown in figure. The influence of ejector spacing on aerodynamic performance in terms of thrust loss due to chute base drag as a function of outer stream pressure ratio is shown in figure. Increasing the ejector spacing results in a marked improvement in the aerodynamic performance because an increased induction of ambient air into the chutes is possible with more spacing. This increased entrainment of ambient air results in improved mixing within the ejector system, and this mixing results in a lower merged velocity and hence lower noise levels.

Figures show the influence of ejector length L on peak PNL and on thrust loss due to base drag. Increased ejector length yields slightly higher noise suppression at all jet velocities because of the containment of sources over a larger distance. The longer ejector shows improved aerodynamic performance as well.

Influence of acoustic treatment attached to the ejector internal surface and/or plug on acoustic characteristics is shown in figure. Successive treatment of the ejector and plug surfaces improves acoustic suppression. The hard-wall ejector provides the suppression by pure physical shielding and ambient air induction. Successive improvements because of the treatment are not very sensitive to treatment impedance.

Suppressor installed on J-79 engine.

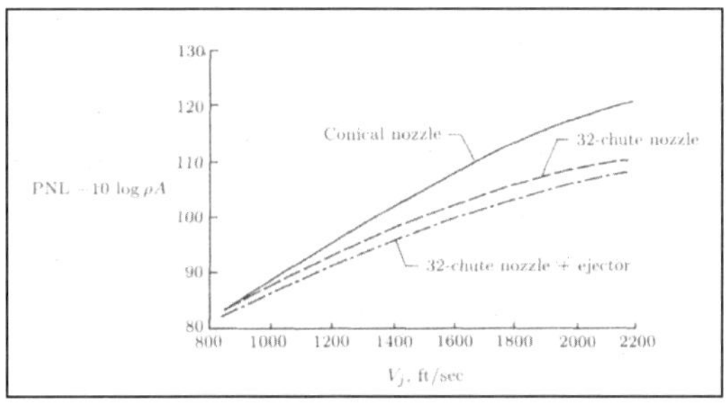

Peak PNL's. 2128-ft sideline; full scale; scale factor, 2:1. General Electric 32-chute annular nozzle on J-79 testbed with and without treated ejector.

Based on these studies, the following design guidelines have evolved:

- An increase in the axial stagger between the suppressor exit plane and the ejector improves acoustic and aerodynamic performance; however, weight and mechanical complexity increase.
- An increase in ejector length improves noise suppression slightly.
- Treatment of the ejector and the plug surface improves noise reduction potential, and the improvement is not sensitive to treatment impedance.

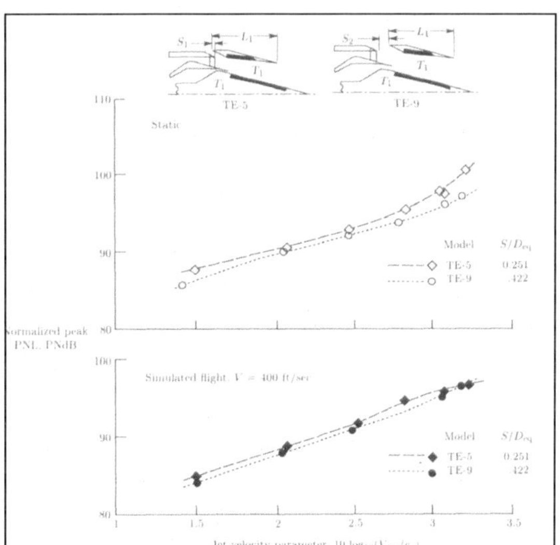

Influence of ejector spacing on normalized peak PNL of 20-shallowchute suppressor nozzle (dual flow). 2400-ft sideline; A_T = 1400 in²; standard day (59 °F, 70 percent relative humidity).

Annular Plug and 2-D Nozzles

Two-dimensional nozzles and plug nozzles with a high radius ratio R_r (i.e., ratio of inner radius to outer radius R_r) exhibit similar flow characteristics near the jet exit plane, since a 2-D nozzle of height h and width b can be viewed as an "unwrapped" annular nozzle of annulus height h and of circumference b. However, the flow characteristics downstream and acoustic characteristics of these two types of nozzles are quite dissimilar.Since they both serve as the first step in departure from the simple conical nozzles to achieve jet noise suppression.

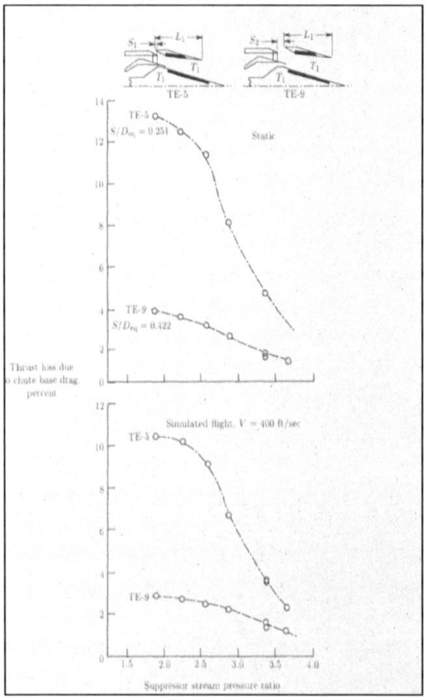

Influence of ejector spacing on base drag of 20-shallow-chute suppressor nozzle (dual flow).

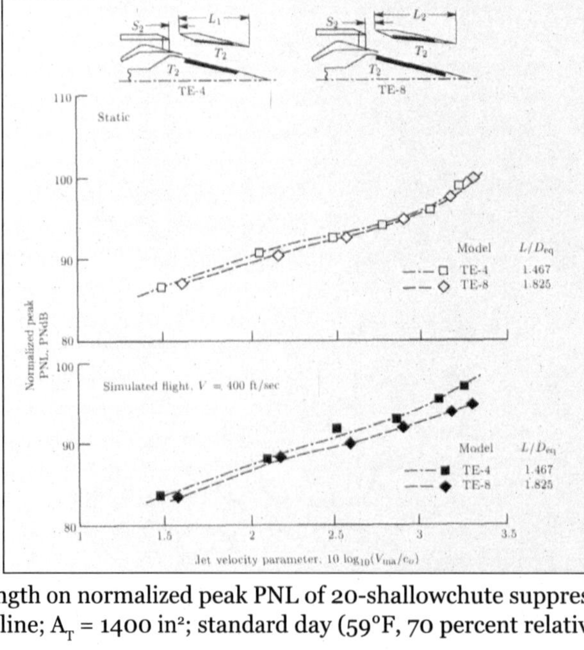

Influence of ejector length on normalized peak PNL of 20-shallowchute suppressor nozzle (dual flow). 2400-ft sideline; A_T = 1400 in²; standard day (59°F, 70 percent relative humidity).

For a given flow area, as the plug nozzle radius ratio increases, annular height decreases and outer radius of nozzle increases. The noise reduction capability of plug nozzles is ascribable to the following:

- Larger surface area is available for shear with ambient air than in the conical nozzle.
- Additional surface (i.e., plug) is available for shearing the jet.
- Smaller characteristic dimension (i.e., annulus height) is present in the highjet-velocity region, and this smaller dimension implies the jet decay and shock structures are governed by annulus height rather than by the equivalent-flowarea circular nozzle diameters close to the exit plane. The annulus height being the characteristic dimension in the high-velocity region results in an acoustic power level spectrum with more high-frequency content than for a circular nozzle. As in the case with multielement suppressors, the high frequency acoustic content suffers larger atmospheric attenuation with plug nozzles than with circular nozzles to yield the observed noise reductions.

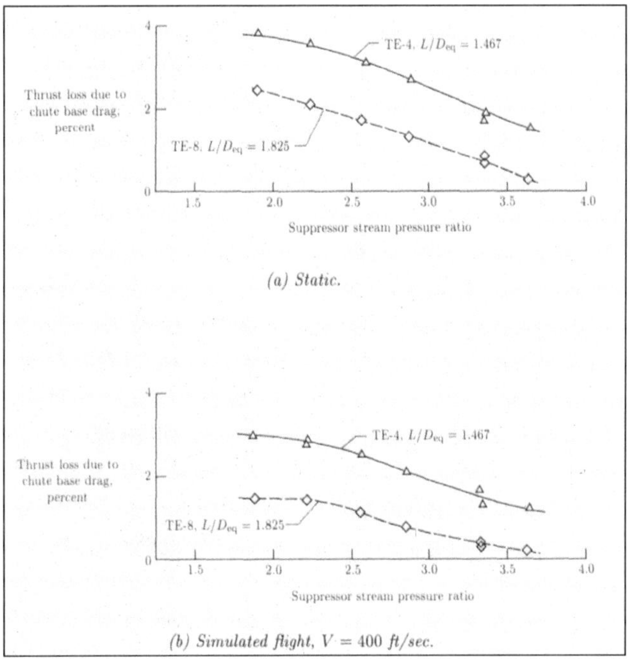

Influence of ejector length on base drag of 20-shallow-chute suppressor nozzle (dual flow).

Figure shows the noise reduction characteristics of annular plug nozzles compared with those of the conical nozzle over a range of jet velocities in the aft quadrant (shown as normalized peak OASPL), which is dominated by jet noise, and in the forward quadrant (shown as normalized 50° OASPL), which is dominated by broadband shock-cell noise. As radius ratio R, increases, the normalized peak OASPL decreases. The trend of shock-cell noise (normalized 50° OASPL) with radius ratio is not very clear. High-radius-ratio plug nozzles which have a blunt tip tend to have two shock structures at high pressure ratios, one on the plug surface (characterized by annulus height) and another downstream of the plug surface due to a supersonic flow expansion fan around the blunt tip. When the downstream shock gets reflected from the jet shear layer it induces a train of shock-cell structures downstream of the plug. The downstream shock structure is typically characterized by the equivalent circular nozzle diameter. Thus, the shock noise contribution observed at $\theta_i = 50°$

consists of contributions from both these two shock structures and a clear trend with radius ratio is absent.

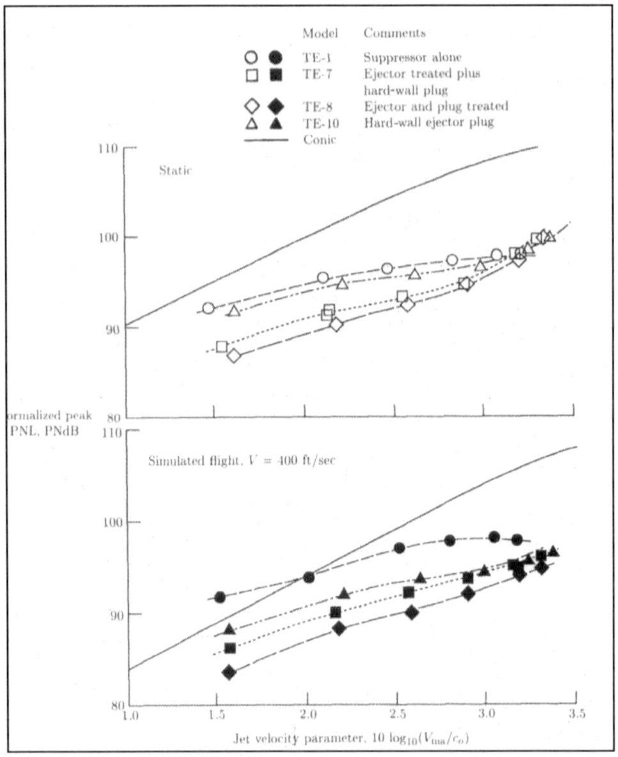

Influence of acoustic treatment a `cached to ejector and/or plug surface on peak PNL of 20-shallow-chute suppressor nozzle (dual flow). 2400-ft side line; A_T = 1400 in^2 ; standard day (59 °F, 70 percent relative humidity).

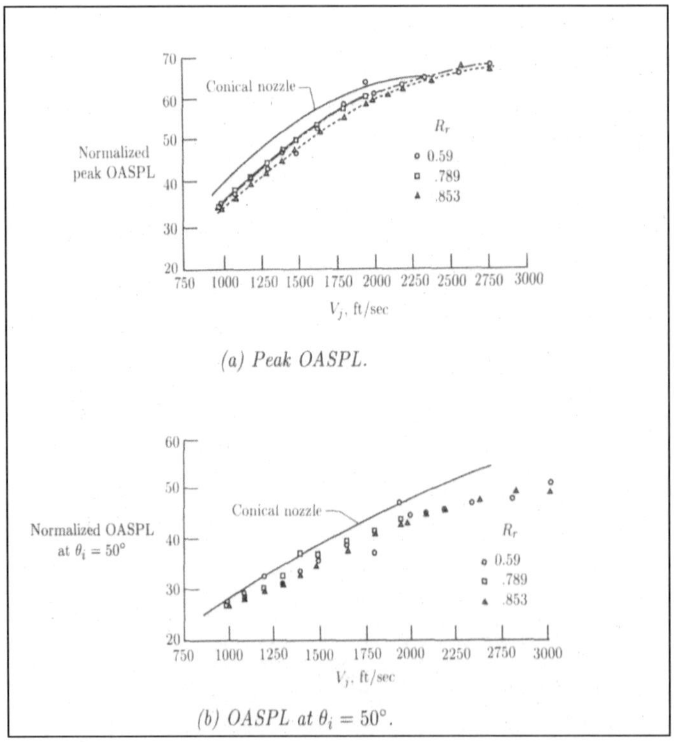

Annular plug nozzle acoustic characteristics as function of radius ratio. 2400-ft sideline; A_T = 338 in^2.

High-radius-ratio plug nozzles have preferred noise characteristics. However, a higher radius ratio for a given flow area dictates a larger nacelle diameter with the accompanying weight and nozzle afterbody drag penalties. A value of 0.853 is considered to be the practical limit for the radius ratio.

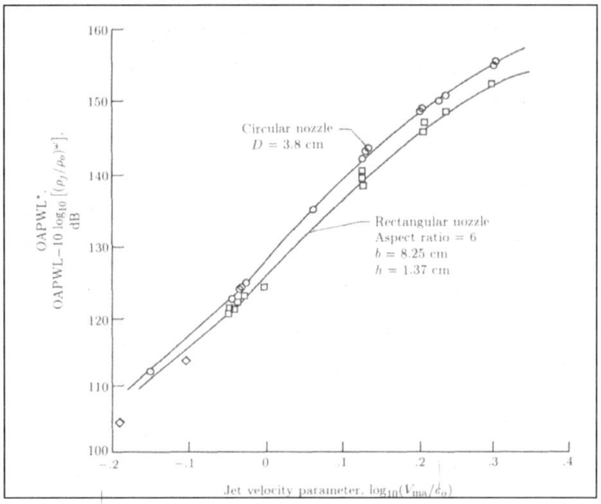

Overall power levels (OAPWL) of circular and rectangular nozzles.

Two-dimensional nozzles (also called rectangular nozzles) are gaining applications in military air-craft because of their thrust-vectoring capability. Two-dimensional nozzles are characterized by their aspect ratios (defined as width/height). At sufficiently high aspect ratios (typically larger than 6), the flow near the exit plane of the jet exhibits a 2-D character. Two-dimensional nozzles tend to be generally quieter than equivalent circular conical nozzles, essentially because of the increased surface area available for shear with the ambient air compared with that of an equivalent circular conical nozzle. Two-dimensional nozzles exhibit azimuthal variation and the plane containing the minor axis is louder than the plane containing the major axis, particularly in the high-frequency region. The 2-D jet flow loses its 2-D nature at downstream locations because of jet diffusion, and at distances sufficiently downstream the flow becomes a circular jet. Since low-frequency noise of jets is associated with large, turbulent eddies and large, turbulent eddies of 2-D nozzles are at regions where the jet is circular, low-frequency noise of 2-D jets typically does not show any azi-muthal variation.

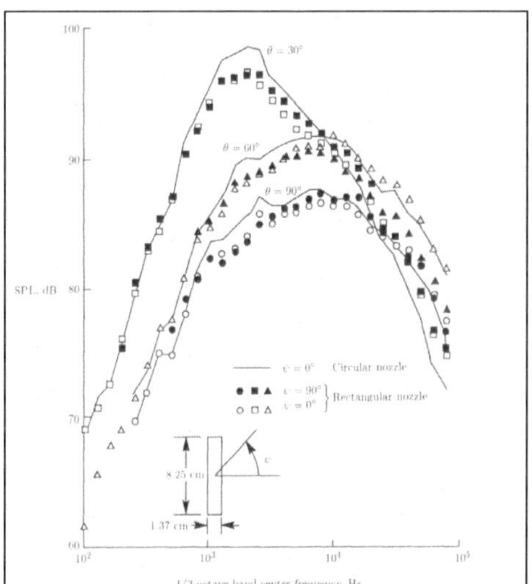

Spectral characteristics of circular and rectangular nozzles. V_i = 966 ft/sec; T_o = 528°R.

Scale-model tests were performed on rectangular jets, and some of the results are presented in figures to justify the statement made above. Increasing the aspect ratio of a 2-D nozzle for a constant flow area results in the jet flow retaining its planar nature for larger distances. Also, increasing the aspect ratio for a given flow area reduces the jet height and thus results in a larger velocity gradient in the vertical direction. This larger velocity gradient in turn yields a faster decay of the jet, although the strength of the high-frequency sources (close to the jet exit plane) increases. Reduction in height also results in smaller shock cells, the structure of which extends to a smaller physical length. The smaller shock cells result in lower shock noise levels. Thus, increasing the aspect ratio of a 2-D nozzle is beneficial acoustically. Mechanical considerations involving airframe-engine integration might limit the aspect ratio.

Gas Power Cycles

CHAPTER 4

Gas power cycles include heat transfer, work transfer, pressure variations, temperature variations, volume variations and entropy variations. There are various gas power cycles namely Brayton cycle, Rankine cycle, Carnot cycle, Diesel cycle, etc. All these different gas power cycles have been carefully analyzed in this chapter.

BRAYTON CYCLE

The Brayton cycle (or Joule cycle) represents the operation of a gas turbine engine. The cycle consists of four processes, as shown in figure alongside a sketch of an engine:

- a - b Adiabatic, quasi-static (or reversible) compression in the inlet and compressor.
- b - c Constant pressure fuel combustion (idealized as constant pressure heat addition).
- c - d Adiabatic, quasi-static (or reversible) expansion in the turbine and exhaust nozzle, with which we:
    - Take some work out of the air and use it to drive the compressor, and
    - Take the remaining work out and use it to accelerate fluid for jet propulsion, or to turn a generator for electrical power generation.
- d - a Cool the air at constant pressure back to its initial condition.

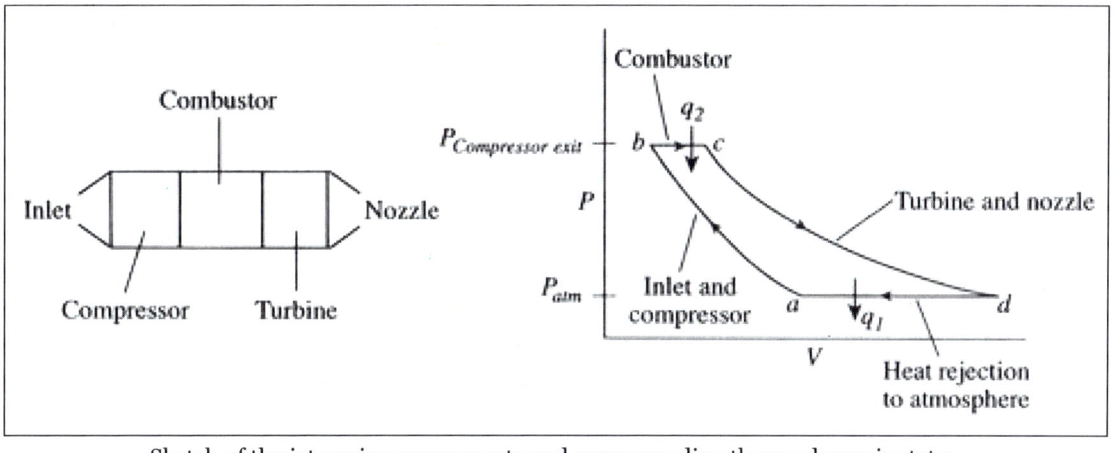

Sketch of the jet engine components and corresponding thermodynamic states.

The components of a Brayton cycle device for jet propulsion are shown in figure. We will typically represent these components schematically, as in figure. In practice, real Brayton cycles take one of two forms. figure shows an "open" cycle, where the working fluid enters and then exits the device. This is the way a jet propulsion cycle works. Figure shows the alternative, a closed cycle, which re-circulates the working fluid. Closed cycles are used, for example, in space power generation.

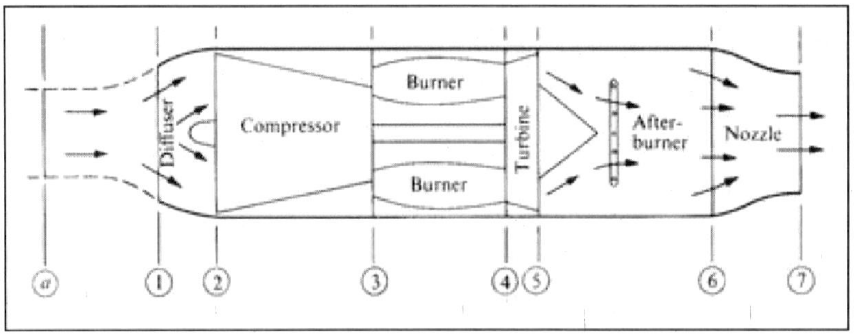

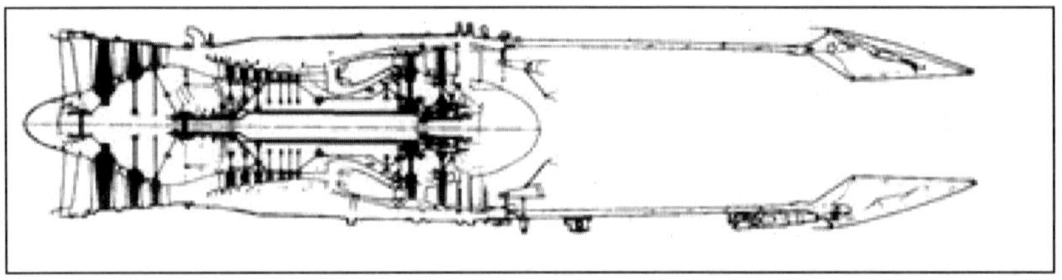

Schematics of typical military gas turbine engines. Top: turbojet with afterburning, bottom: GE F404 low bypass ratio turbofan with afterburning.

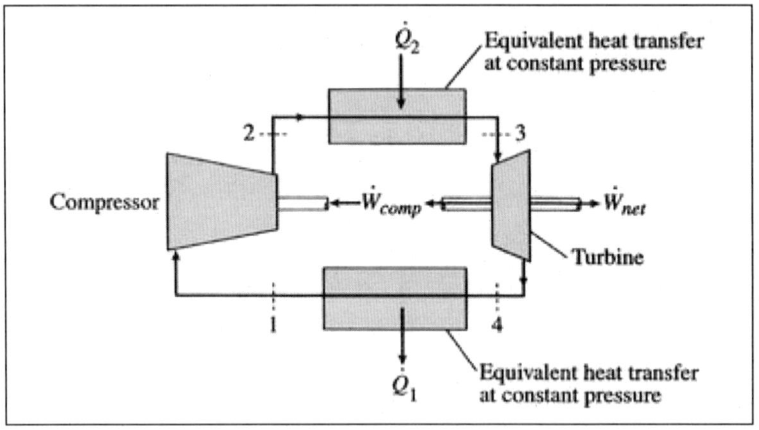

Thermodynamic model of gas turbine engine cycle for power generation.

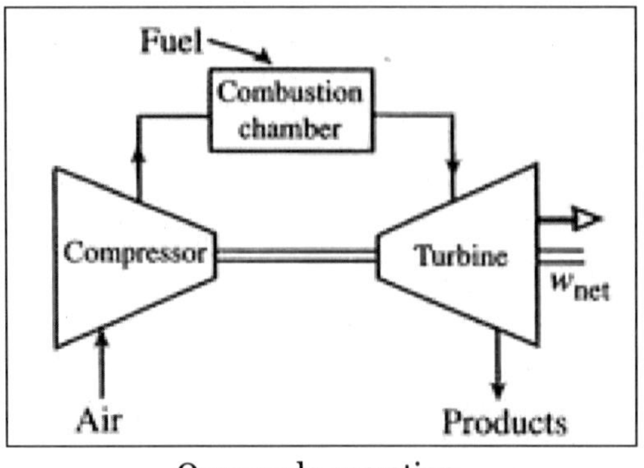

Open cycle operation.

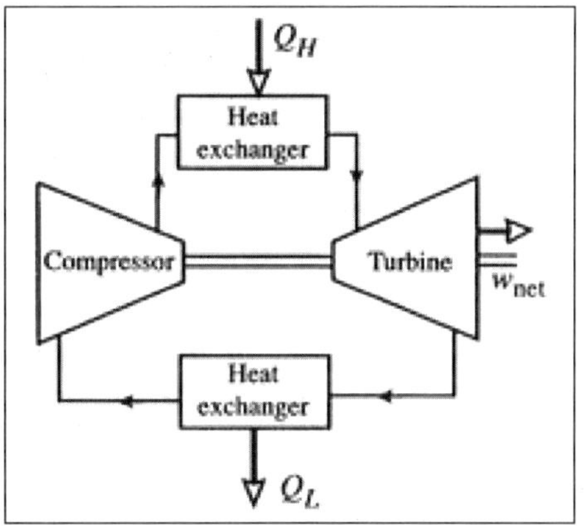

Closed cycle operation, Options for operating Brayton cycle gas turbine engines.

Work and Efficiency

The objective now is to find the work done, the heat absorbed, and the thermal efficiency of the cycle. Tracing the path shown around the cycle from $a-b-c-d$ and back to a , the first law gives (writing the equation in terms of a unit mass),

$$\Delta u_{a\text{-}b\text{-}c\text{-}d\text{-}a} = 0 = q_2 + q_1 - w.$$

Here Δu is zero because u is a function of state, and any cycle returns the system to its starting state. The net work done is therefore,

$$w = q_2 + q_1,$$

where q_1, q_2 are defined as heat received by the system (q_1 is negative). We thus need to evaluate the heat transferred in processes $b-c$ and d - a.

For a constant pressure, quasi-static process the heat exchange per unit mass is:

$$dh = c_p dT = dq, \quad \text{or} \quad [dq]_{\text{constant P}} = dh.$$

We can see this by writing the first law in terms of enthalpy or by remembering the definition of c_p

The heat exchange can be expressed in terms of enthalpy differences between the relevant states. Treating the working fluid as a perfect gas with constant specific heats, for the heat addition from the combustor,

$$q_2 = h_c - h_b = c_p(T_c - T_b).$$

The heat rejected is, similarly,

$$q_1 = h_a - h_d = c_p(T_a - T_d).$$

The network per unit mass is given by:

$$\text{Net work per unit mass} = q_1 + q_2 = c_p[(T_c - T_b) + (T_a - T_d)].$$

The thermal efficiency of the Brayton cycle can now be expressed in terms of the temperatures:

$$\eta = \frac{\text{Net work}}{\text{Heat in}} = \frac{c_p[(T_c - T_b) - (T_d - T_a)]}{c_p(T_c - T_b)} = 1 - \frac{(T_d - T_a)}{(T_c - T_b)} = 1 - \frac{T_a(T_d / T_a - 1)}{T_b(T_c / T_b - 1)}.$$

To proceed further, we need to examine the relationships between the different temperatures. We know that points a and d are on a constant pressure process as are points b and c, and $P_a = P_d; P_b = P_c$.. The other two legs of the cycle are adiabatic and reversible, so:

$$\frac{P_d}{P_c} = \frac{P_a}{P_b} \quad \Rightarrow \quad \left(\frac{T_d}{T_c}\right)^{\gamma/(\gamma-1)} = \left(\frac{T_a}{T_b}\right)^{\gamma/(\gamma-1)}.$$

Therefore $T_d / T_c = T_a / T_b$, or, finally, $T_d / T_a = T_c / T_b$.. Using this relation in the expression for

thermal efficiency, Eq. $\eta = \frac{\text{Net work}}{\text{Heat in}} = \frac{c_p[(T_c - T_b) - (T_d - T_a)]}{c_p(T_c - T_b)} = 1 - \frac{(T_d - T_a)}{(T_c - T_b)} = 1 - \frac{T_a(T_d / T_a - 1)}{T_b(T_c / T_b - 1)}.$

yields an expression for the thermal efficiency of a Brayton cycle:

$$\text{Ideal Brayton cycle efficiency}: \eta_B = 1 - \frac{T_a}{T_b} = 1 - \frac{T_{\text{atmospheric}}}{T_{\text{compressor exit}}}.$$

The temperature ratio across the compressor, $T_b / T_a = TR$. In terms of compressor temperature ratio, and using the relation for an adiabatic reversible process we can write the efficiency in terms of the compressor (and cycle) pressure ratio, which is the parameter commonly used:

$$\eta_B = 1 - \frac{1}{TR} = 1 - \frac{1}{PR^{(\gamma-1)/\gamma}}.$$

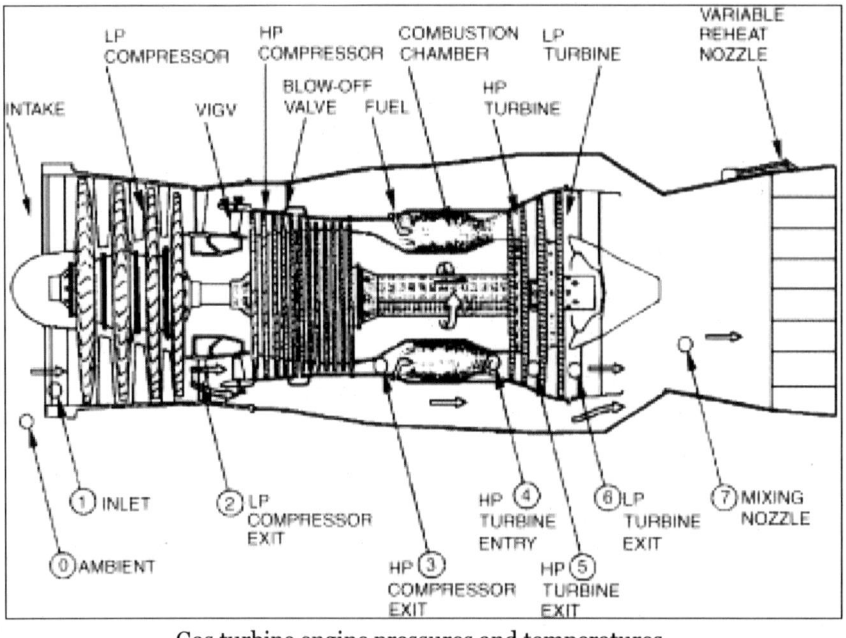

Gas turbine engine pressures and temperatures.

Figure shows pressures and temperatures through a gas turbine engine (the PW4000, which powers the 747 and the 767).

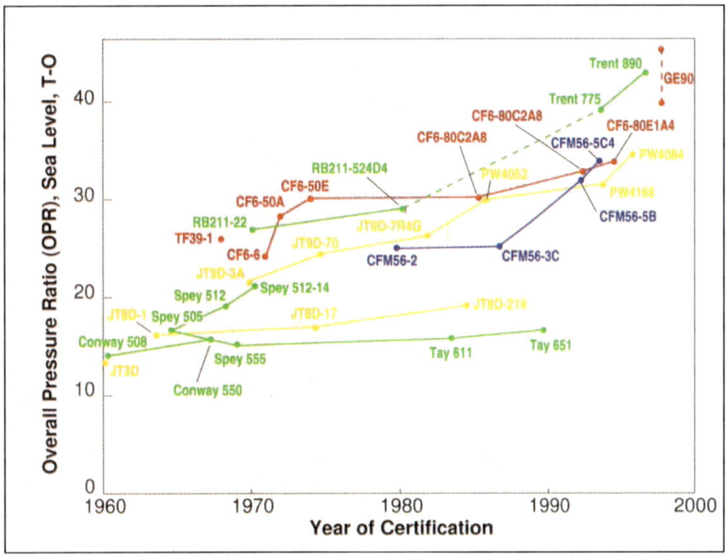

Gas turbine engine pressure ratio trends.

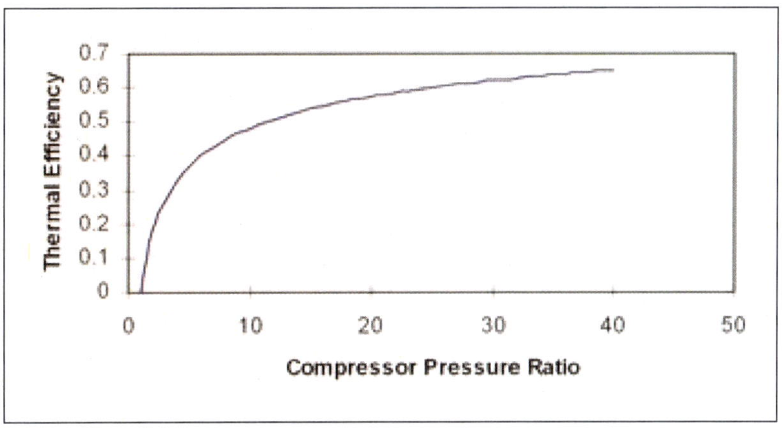

Trend of Brayton cycle thermal efficiency with compressor pressure ratio.

Equation $\eta_B = 1 - \frac{1}{TR} = 1 - \frac{1}{PR^{(\gamma-1)/\gamma}}$. says that for a high cycle efficiency, the pressure ratio of the cycle should be increased. This trend is plotted in figure. Figure shows the history of aircraft engine pressure ratio versus entry into service, and it can be seen that there has been a large increase in cycle pressure ratio. The thermodynamic concepts apply to the behavior of real aerospace devices.

Gas Turbine Technology and Thermodynamics

The turbine entry temperature, T_c, is fixed by materials technology and cost. (If the temperature is too high, the blades fail.) figures and show the progression of the turbine entry temperatures in aeroengines. Figure is from Rolls Royce and figure is from Pratt & Whitney. Note the relation between the gas temperature coming into the turbine blades and the blade melting temperature.

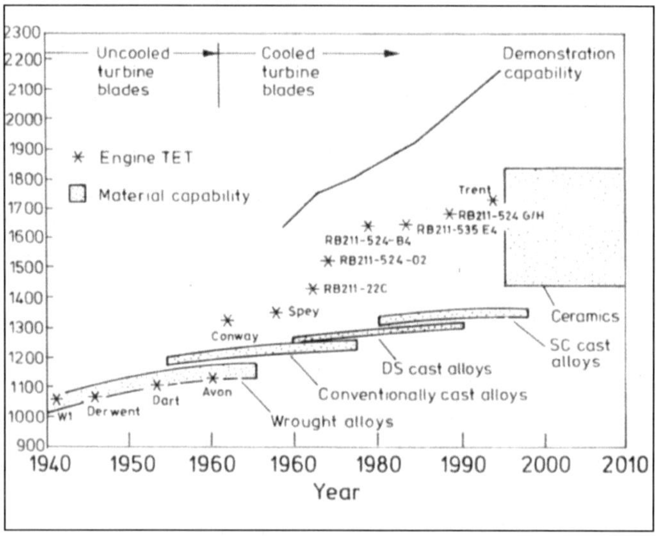

Rolls-Royce high temperature technology.

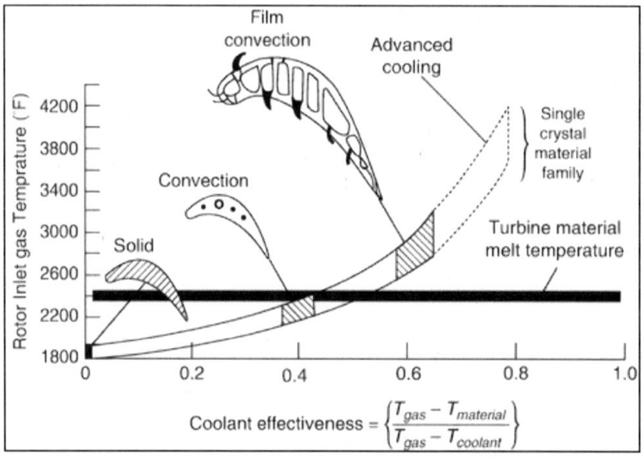

Turbine blade cooling technology [Pratt & Whitney].

For a given level of turbine technology (in other words given maximum temperature) a design question is what should the compressor *TR* be? What criterion should be used to decide this? Maximum thermal efficiency? Maximum work? We examine this issue below.

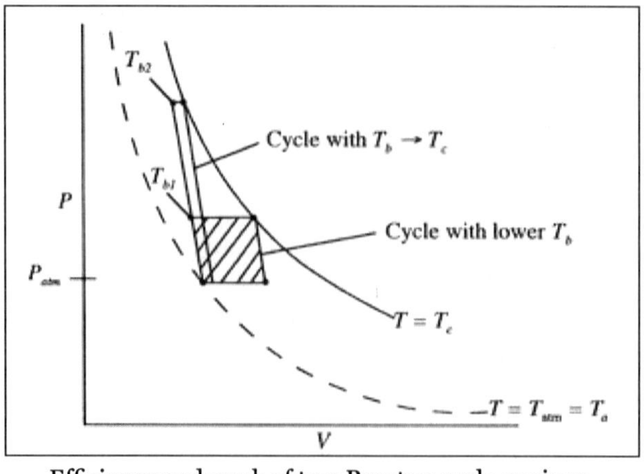

Efficiency and work of two Brayton cycle engines.

The problem is posed in figure, which shows two Brayton cycles. For maximum efficiency we would like *TR* as high as possible. This means that the compressor exit temperature approaches the turbine entry temperature. The net work will be less than the heat received; as $T_b \rightarrow T_c$ the heat received approaches zero and so does the net work.

The net work in the cycle can also be expressed as $\int Pdv$, evaluated in traversing the cycle. This is the area enclosed by the curves, which is seen to approach zero as $T_b \rightarrow T_c$.

The conclusion from either of these arguments is that a cycle designed for maximum thermal efficiency is not very useful in that the work (power) we get out of it is zero.

A more useful criterion is that of maximum work per unit mass (maximum power per unit mass flow). This leads to compact propulsion devices. The work per unit mass is given by:

$$\text{Work/unit mass} = c_p[(T_c - T_b) - (T_d - T_a)],$$

where T_c is the maximum turbine inlet temperature (a design constraint) and T_a is atmospheric temperature. The design variable is the compressor exit temperature, T_b, and to find the maximum as this is varied, we differentiate the expression for work with respect to T_b:

$$\frac{d\text{Work}}{dT_b} = c_p\left[\frac{dT_c}{dT_b} - 1 - \frac{dT_d}{dT_b} + \frac{dT_a}{dT_b}\right].$$

The first and the fourth terms on the right hand side of the above equation are both zero (the turbine entry temperature is fixed, as is the atmospheric temperature). The maximum work occurs where the derivative of work with respect to T_b is zero:

$$\frac{d\text{Work}}{dT_b} = 0 = -1 - \frac{dT_d}{dT_b}.$$

To use Eq. $\frac{d\text{Work}}{dT_b} = 0 = -1 - \frac{dT_d}{dT_b}.$, we need to relate T_d and T_b. We know that:

$$\frac{T_d}{T_a} = \frac{T_c}{T_b} \quad \text{or} \quad T_d = \frac{T_a T_c}{T_b}.$$

Hence,

$$\frac{dT_d}{dT_b} = -\frac{T_a T_c}{T_b^2}.$$

Plugging this expression for the derivative into Eq. $\frac{d\text{Work}}{dT_b} = 0 = -1 - \frac{dT_d}{dT_b}.$ gives the compressor exit temperature for maximum work as $T_b = \sqrt{T_a T_c}$.. In terms of temperature ratio,

$$\text{Compressor temperature ratio for maximum work}: \frac{T_b}{T_a} = \sqrt{\frac{T_c}{T_a}}.$$

The condition for maximum work in a Brayton cycle is different than that for maximum efficiency. The role of the temperature ratio can be seen if we examine the work per unit mass which is delivered at this condition:

$$\text{Work/unit mass} = c_p\left[T_c - \sqrt{T_aT_c} - \frac{T_aT_c}{\sqrt{T_aT_c}} + T_a\right].$$

Ratioing all temperatures to the engine inlet temperature,

$$\text{Work/unit mass} = c_pT_a\left[\frac{T_c}{T_a} - 2\sqrt{\frac{T_c}{T_a}} + 1\right].$$

To find the power the engine can produce, we need to multiply the work per unit mass by the mass flow rate:

$$\text{Power} = \dot{m}c_pT_a\left[\frac{T_c}{T_a} - 2\sqrt{\frac{T_c}{T_a}} + 1\right]; \text{Maximum power for an ideal Brayton cycle.}$$

(The units are $\frac{kg}{s}\frac{J}{kg\text{-}K}K = \frac{J}{s} = \text{Watts.}$)

The trend of work output vs. compressor pressure ratio, for different temperature ratios $\text{TR} = T_c / T_a$, is shown in figure.

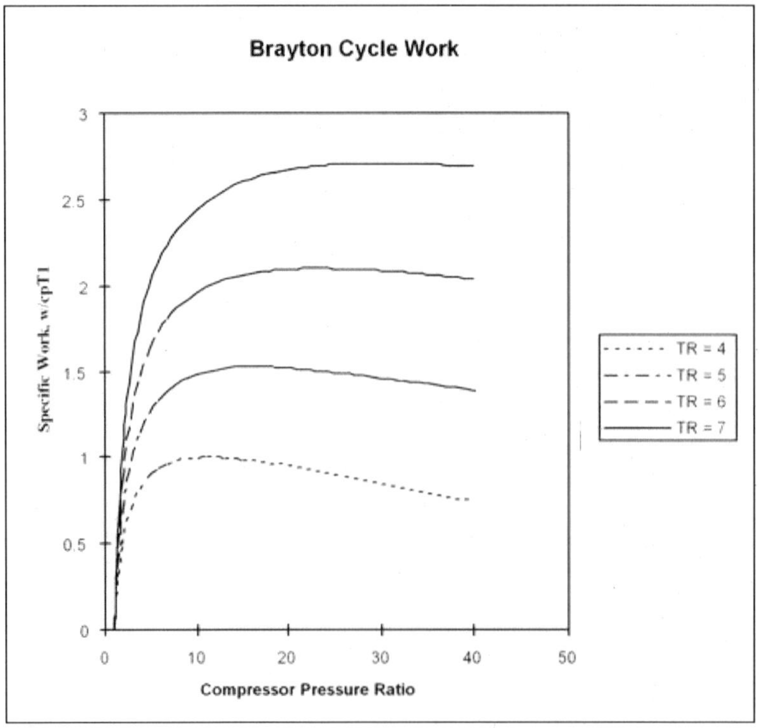

Trend of cycle work with compressor pressure ratio, for different temperature ratios $\text{TR} = T_c / T_a$.

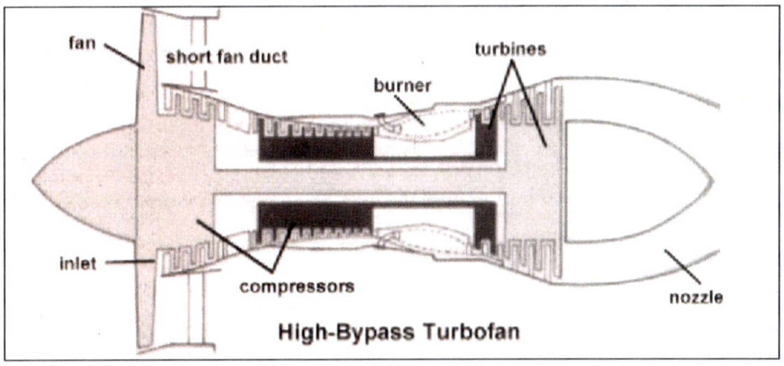

Gas turbine engine core.

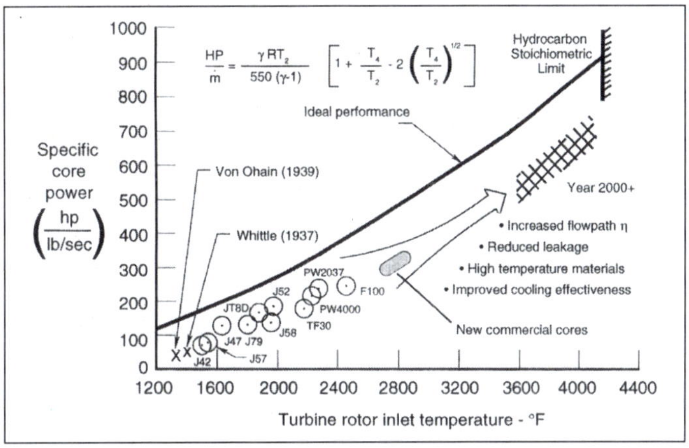

Core power vs. turbine entry temperature
Aeroengine core power.

Figure shows the expression for power of an ideal cycle compared with data from actual jet engines. Figure shows the gas turbine engine layout including the core (compressor, burner, and turbine). Figure shows the core power for a number of different engines as a function of the turbine rotor entry temperature. The equation in the figure for horsepower (HP) is the same as that which we just derived, except for the conversion factors. The analysis not only shows the qualitative trend very well but captures much of the quantitative behavior too.

A final comment on Brayton cycles concerns the value of the thermal efficiency. The Brayton cycle thermal efficiency contains the ratio of the compressor exit temperature to atmospheric temperature, so that the ratio is not based on the highest temperature in the cycle, as the Carnot efficiency is. For a given maximum cycle temperature, the Brayton cycle is therefore less efficient than a Carnot cycle.

Brayton Cycle for Jet Propulsion: The Ideal Ramjet

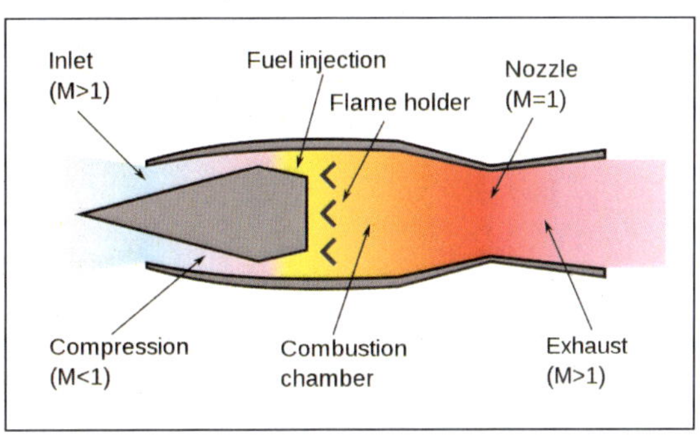

Ideal ramjet.

In the ramjet there are "no moving parts." The processes that occur in this propulsion device are:

- $0 \rightarrow 3$: Isentropic diffusion (slowing down) and compression, with a decrease in Mach number, $M_0 \rightarrow M_3 \ll 1$.
- $3 \rightarrow 4$: Constant pressure combustion.
- $4 \rightarrow 5$: Isentropic expansion through the nozzle.

The ramjet thermodynamic cycle efficiency can be written in terms of flight Mach number, M_0, as follows:

$$\eta_{\text{Brayton}} = 1 - \frac{T_0}{T_{\text{compressor exit}}} = 1 - \frac{T_0}{T_3} = 1 - \frac{T_0}{T_{t0}}$$

and,

$$\frac{T_0}{T_{t0}} = \frac{1}{1 + \frac{\gamma - 1}{2} M_0^2},$$

so,

$$\eta_{\text{Brayton}} = \frac{\frac{\gamma - 1}{2} M_0^2}{1 + \frac{\gamma - 1}{2} M_0^2}.$$

RANKINE CYCLE

The Rankine cycle is a model used to predict the performance of steam turbine systems. It was also used to study the performance of reciprocating steam engines. The Rankine cycle is an idealized thermodynamic cycle of a heat engine that converts heat into mechanical work while undergoing phase change. It is an idealized cycle in which friction losses in each of the four components are neglected. The heat is supplied externally to a closed loop, which usually uses water as the working fluid. It is named after William John Macquorn Rankine, a Scottish polymath and Glasgow University professor.

The Rankine cycle closely describes the process by which steam-operated heat engines commonly found in thermal power generation plants generate power.

Power depends on the temperature difference between a heat source and a cold source. The higher the difference, the more mechanical power can be efficiently extracted out of heat energy, as per Carnot's theorem.

The heat sources used in these power plants are usually nuclear fission or the combustion of fossil fuels such as coal, natural gas, and oil, or concentrated solar power. The higher the temperature, the better.

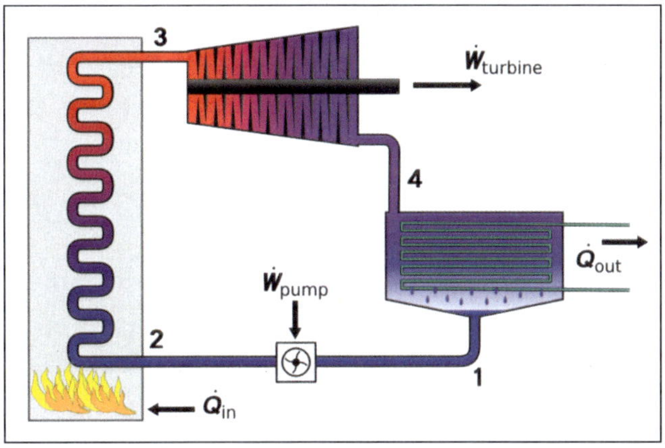

Physical layout of the Rankine cycle 1. Pump, 2. Boiler, 3. Turbine, 4. Condenser.

The efficiency of the Rankine cycle is limited by the high heat of vaporization of the working fluid. Also, unless the pressure and temperature reach super critical levels in the steam boiler, the temperature range the cycle can operate over is quite small: steam turbine entry temperatures are typically around 565 °C and steam condenser temperatures are around 30 °C. This gives a theoretical maximum Carnot efficiency for the steam turbine alone of about 63.8% compared with an actual overall thermal efficiency of up to 42% for a modern coal-fired power station. This low steam turbine entry temperature (compared to a gas turbine) is why the Rankine (steam) cycle is often used as a bottoming cycle to recover otherwise rejected heat in combined-cycle gas turbine power stations.

The cold source (the colder the better) used in these power plants are usually cooling towers and a large water body (river or sea). The efficiency of the Rankine cycle is limited on the cold side by the lower practical temperature of the working fluid.

The working fluid in a Rankine cycle follows a closed loop and is reused constantly. The water vapor with condensed droplets often seen billowing from power stations is created by the cooling systems (not directly from the closed-loop Rankine power cycle). This 'exhaust' heat is represented by the "Q_{out}" flowing out of the lower side of the cycle shown in the T–s diagram below. Cooling towers operate as large heat exchangers by absorbing the latent heat of vaporization of the working fluid and simultaneously evaporating cooling water to the atmosphere.

While many substances could be used as the working fluid in the Rankine cycle, water is usually the fluid of choice due to its favorable properties, such as its non-toxic and unreactive chemistry, abundance, and low cost, as well as its thermodynamic properties. By condensing the working steam vapor to a liquid the pressure at the turbine outlet is lowered and the energy required by the feed pump consumes only 1% to 3% of the turbine output power and these factors contribute to a higher efficiency for the cycle. The benefit of this is offset by the low temperatures of steam admitted to the turbines. Gas turbines, for instance, have turbine entry temperatures approaching 1500 °C. However, the thermal efficiency of actual large steam power stations and large modern gas turbine stations are similar.

The Four Processes in the Rankine Cycle

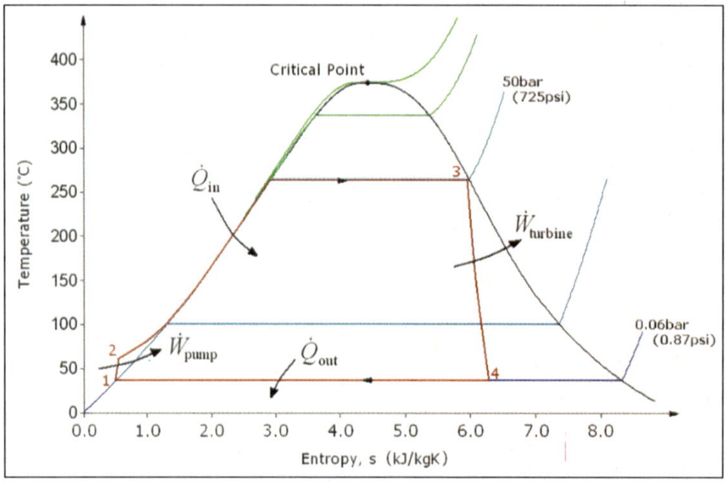

T–s diagram of a typical Rankine cycle operating between pressures of 0.06 bar and 50 bar.
Left from the bell-shaped curve is liquid, right from it is gas, and under it is saturated liquid–vapour equilibrium.

There are four processes in the Rankine cycle. The states are identified by numbers (in brown) in the T–s diagram.

- Process 1–2: The working fluid is pumped from low to high pressure. As the fluid is a liquid at this stage, the pump requires little input energy.

In other words Process 1-2 is [Isentropic compression in pump].

- Process 2–3: The high-pressure liquid enters a boiler, where it is heated at constant pressure by an external heat source to become a dry saturated vapour. The input energy required can be easily calculated graphically, using an enthalpy–entropy chart (h–s chart, or Mollier diagram), or numerically, using steam tables.

In other words Process 2-3 is [Constant pressure heat addition in boiler].

- Process 3–4: The dry saturated vapour expands through a turbine, generating power. This decreases the temperature and pressure of the vapour, and some condensation may occur. The output in this process can be easily calculated using the chart or tables noted above.

In other words Process 3-4 is [Isentropic expansion in turbine].

- Process 4–1: The wet vapour then enters a condenser, where it is condensed at a constant pressure to become a saturated liquid.

In other words Process 4-1 is [Constant pressure heat rejection in condenser].

In an ideal Rankine cycle the pump and turbine would be isentropic, i.e., the pump and turbine would generate no entropy and hence maximize the net work output. Processes 1–2 and 3–4 would be represented by vertical lines on the T–s diagram and more closely resemble that of the Carnot cycle. The Rankine cycle shown here prevents the state of the working fluid from ending up in the superheated vapor region after the expansion in the turbine,which reduces the energy removed by the condensers.

The actual vapor power cycle differs from the ideal Rankine cycle because of irreversibilities in the inherent components caused by fluid friction and heat loss to the surroundings; fluid friction causes pressure drops in the boiler, the condenser, and the piping between the components, and as a result the steam leaves the boiler at a lower pressure; heat loss reduces the net work output, thus heat addition to the steam in the boiler is required to maintain the same level of net work output.

Variables

$\dot{Q}$	Heat flow rate to or from the system (energy per unit time)
$\dot{m}$	Mass flow rate (mass per unit time)
$\dot{W}$	Mechanical power consumed by or provided to the system (energy per unit time)
η_{therm}	Thermodynamic efficiency of the process (net power output per heat input, dimensionless)
$\eta_{\text{pump}}, \eta_{\text{turb}}$	Isentropic efficiency of the compression (feed pump) and expansion (turbine) processes, dimensionless
h_1, h_2, h_3, h_4	The "specific enthalpies" at indicated points on the T–s diagram
h_{4s}	The final "specific enthalpy" of the fluid if the turbine were isentropic
p_1, p_2	The pressures before and after the compression process

Equations

In general, the efficiency of a simple rankine cycle can be written as

$$\eta_{\text{therm}} = \frac{\dot{W}_{\text{thermal}} - \dot{W}}{\dot{Q}_{\text{in}}} \approx \frac{\dot{W}_{\text{turb}}}{\dot{Q}_{\text{in}}}.$$

Each of the next four equations is derived from the energy and mass balance for a control volume. η_{therm} defines the thermodynamic efficiency of the cycle as the ratio of net power output to heat input. As the work required by the pump is often around 1% of the turbine work output, it can be simplified.

$$\frac{\dot{Q}_{\text{in}}}{\dot{m}} = h_3 - h_2,$$

$$\frac{\dot{Q}_{\text{out}}}{\dot{m}} = h_4 - h_1,$$

$$\frac{\dot{W}_{\text{pump}}}{\dot{m}} = h_2 - h_1,$$

$$\frac{\dot{W}_{\text{turbine}}}{\dot{m}} = h_3 - h_4.$$

When dealing with the efficiencies of the turbines and pumps, an adjustment to the work terms must be made:

$$\frac{\dot{W}_{\text{pump}}}{\dot{m}} = h_2 - h_1 \approx \frac{v_1 \Delta p}{\eta_{\text{pump}}} = \frac{v_1 (p_2 - p_1)}{\eta_{\text{pump}}},$$

$$\frac{\dot{W}_{\text{turbine}}}{\dot{m}} = h_3 - h_4 \approx (h_3 - h_4)\eta_{\text{turbine}}.$$

Real Rankine Cycle (Non-ideal)

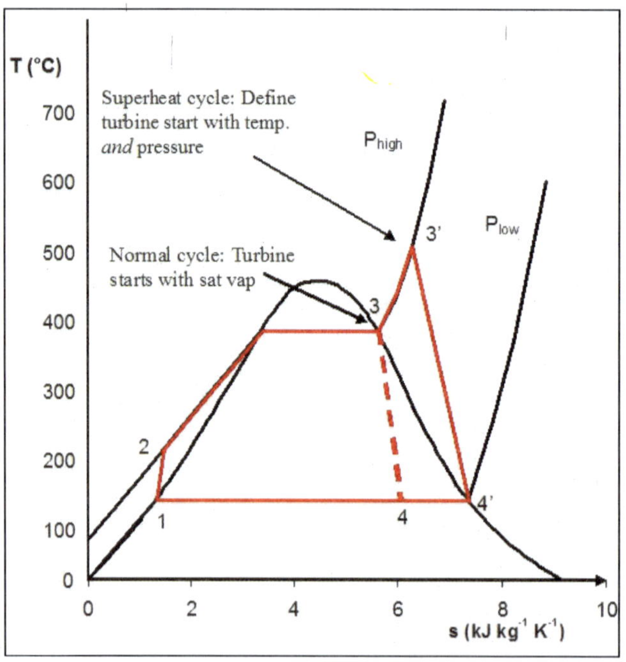

Rankine cycle with superheat.

In a real power-plant cycle (the name "Rankine" cycle is used only for the ideal cycle), the compression by the pump and the expansion in the turbine are not isentropic. In other words, these processes are non-reversible, and entropy is increased during the two processes. This somewhat increases the power required by the pump and decreases the power generated by the turbine.

In particular, the efficiency of the steam turbine will be limited by water-droplet formation. As the water condenses, water droplets hit the turbine blades at high speed, causing pitting and erosion, gradually decreasing the life of turbine blades and efficiency of the turbine. The easiest way to overcome this problem is by superheating the steam. On the T–s diagram above, state 3 is at a border of the two-phase region of steam and water, so after expansion the steam will be very wet. By superheating, state 3 will move to the right (and up) in the diagram and hence produce a drier steam after expansion.

Variations of the Basic Rankine Cycle

The overall thermodynamic efficiency can be increased by raising the average heat input temperature of that cycle. Increasing the temperature of the steam into the superheat region is a simple way of doing this. There are also variations of the basic Rankine cycle designed to raise the thermal efficiency of the cycle in this way; two of these are described below.

$$\overline{T}_{in} = \frac{\int_2^3 T\,dQ}{Q_{in}}$$

Rankine Cycle with Reheat

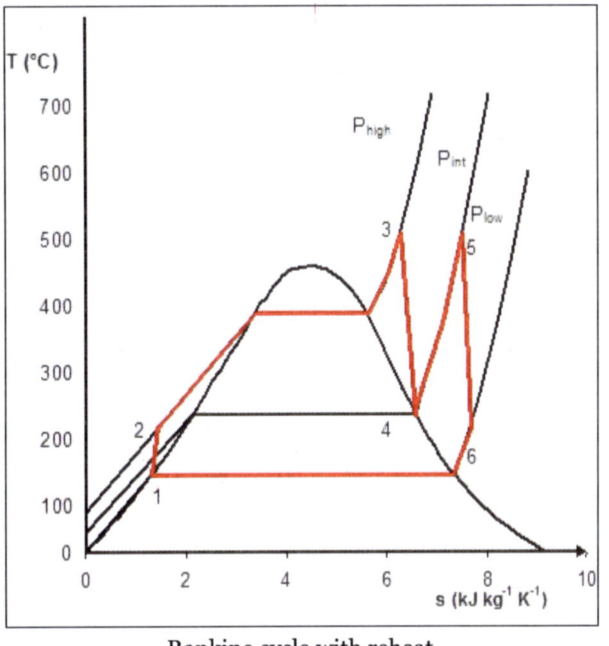

Rankine cycle with reheat.

The purpose of a reheating cycle is to remove the moisture carried by the steam at the final stages of the expansion process. In this variation, two turbines work in series. The first accepts vapor from the boiler at high pressure. After the vapor has passed through the first turbine, it re-enters the boiler and is reheated before passing through a second, lower-pressure, turbine. The reheat temperatures are very close or equal to the inlet temperatures, whereas the optimal reheat pressure needed is only one fourth of the original boiler pressure. Among other advantages, this prevents the vapor from condensing during its expansion and thereby reducing the damage in the turbine blades, and improves the efficiency of the cycle, because more of the heat flow into the cycle occurs at higher temperature. The reheat cycle was first introduced in the 1920s, but was not operational for long due to technical difficulties. In the 1940s, it was reintroduced with the increasing manufacture of high-pressure boilers, and eventually double reheating was introduced in the 1950s. The idea behind double reheating is to increase the average temperature. It was observed that more than two stages of reheating are generally unnecessary, since the next stage increases the cycle efficiency only half as much as the preceding stage. Today, double reheating is commonly used in power plants that operate under supercritical pressure.

Regenerative Rankine Cycle

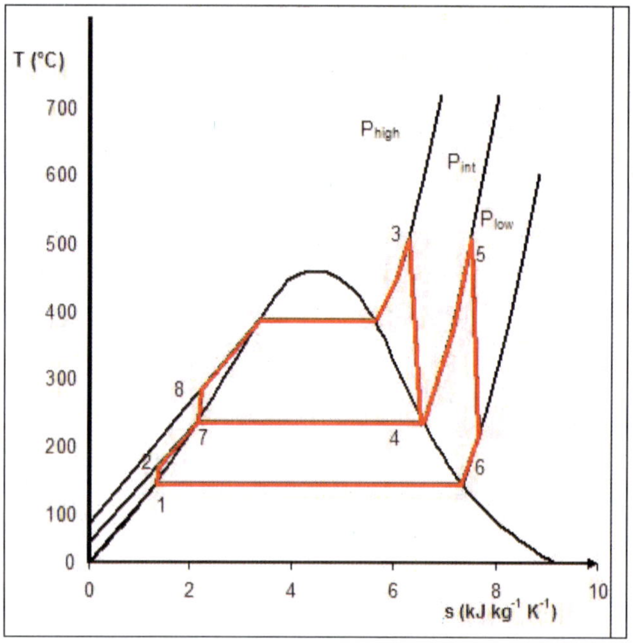

Regenerative Rankine cycle.

The regenerative Rankine cycle is so named because after emerging from the condenser (possibly as a subcooled liquid) the working fluid is heated by steam tapped from the hot portion of the cycle. On the diagram shown, the fluid at 2 is mixed with the fluid at 4 (both at the same pressure) to end up with the saturated liquid at 7. This is called "direct-contact heating". The Regenerative Rankine cycle (with minor variants) is commonly used in real power stations.

Another variation sends bleed steam from between turbine stages to feedwater heaters to preheat the water on its way from the condenser to the boiler. These heaters do not mix the input steam and condensate, function as an ordinary tubular heat exchanger, and are named "closed feedwater heaters".

Regeneration increases the cycle heat input temperature by eliminating the addition of heat from the boiler/fuel source at the relatively low feedwater temperatures that would exist without regenerative feedwater heating. This improves the efficiency of the cycle, as more of the heat flow into the cycle occurs at higher temperature.

Organic Rankine Cycle

The Organic Rankine Cycle (ORC) uses an organic fluid such as n-pentane or toluene in place of water and steam. This allows use of lower-temperature heat sources, such as solar ponds, which typically operate at around 70–90 °C. The efficiency of the cycle is much lower as a result of the lower temperature range, but this can be worthwhile because of the lower cost involved in gathering heat at this lower temperature. Alternatively, fluids can be used that have boiling points above water, and this may have thermodynamic benefits. The properties of the actual working fluid has great influence on the quality of steam (vapour) after the expansion step, influencing the design of the whole cycle.

The Rankine cycle does not restrict the working fluid in its definition, so the name "organic cycle" is simply a marketing concept and the cycle should not be regarded as a separate thermodynamic cycle.

Supercritical Rankine Cycle

The Rankine cycle applied using a supercritical fluid combines the concepts of heat regeneration and supercritical Rankine cycle into a unified process called the regenerative supercritical cycle (RGSC). It is optimised for temperature sources 125–450 °C.

OTTO CYCLE

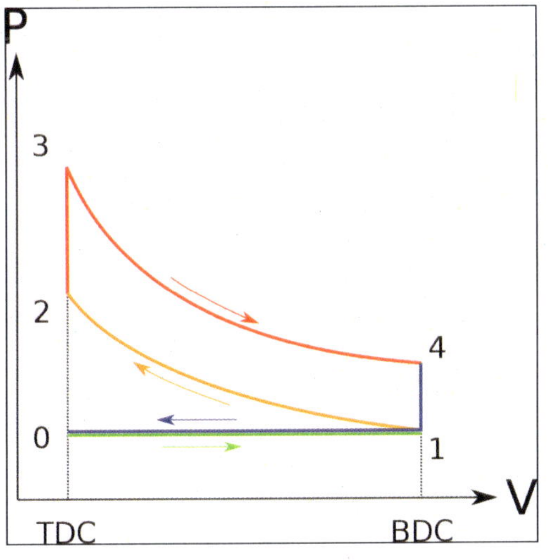

Pressure–volume diagram.

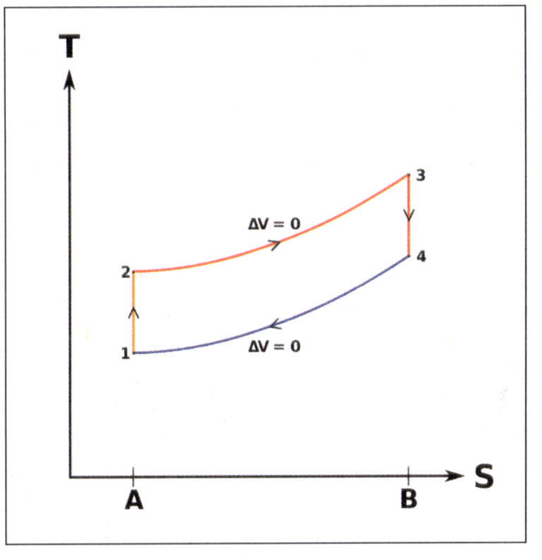

Temperature-Entropy diagram.

The idealized diagrams of a four-stroke Otto cycle Both diagrams: the intake (*A*) stroke is performed by an isobaric expansion, followed by an adiabatic compression (*B*) stroke. Through the combustion of fuel, heat is added in a constant volume (isochoric process) process, followed by an adiabatic expansion process power (*C*) stroke. The cycle is closed by the exhaust (*D*) stroke, characterized by isochoric cooling and isentropic compression processes.

An Otto cycle is an idealized thermodynamic cycle that describes the functioning of a typical spark ignition piston engine. It is the thermodynamic cycle most commonly found in automobile engines.

The Otto cycle is a description of what happens to a mass of gas as it is subjected to changes of pressure, temperature, volume, addition of heat, and removal of heat. The mass of gas that is subjected to those changes is called the system. The system, in this case, is defined to be the fluid (gas) within the cylinder. By describing the changes that take place within the system, it will also describe in inverse, the system's effect on the environment. In the case of the Otto cycle, the effect will be to produce enough net work from the system so as to propel an automobile and its occupants in the environment.

The Otto cycle is constructed from:

- Top and bottom of the loop: a pair of quasi-parallel and isentropic processes (frictionless, adiabatic reversible).
- Left and right sides of the loop: a pair of parallel isochoric processes (constant volume).

The isentropic process of compression or expansion implies that there will be no inefficiency (loss of mechanical energy), and there be no transfer of heat into or out of the system during that process. The cylinder and piston are assumed to be impermeable to heat during that time. Work is performed on the system during the lower isentropic compression process. Heat flows into the Otto cycle through the left pressurizing process and some of it flows back out through the right depressurizing process. The summation of the work added to the system plus the heat added minus the heat removed yields the net mechanical work generated by the system.

Processes

The processes are described by:

- Process 0–1 a mass of air is drawn into piston/cylinder arrangement at constant pressure.
- Process 1–2 is an adiabatic (isentropic) compression of the charge as the piston moves from bottom dead centre (*BDC*) to top dead centre (*TDC*).
- Process 2–3 is a constant-volume heat transfer to the working gas from an external source while the piston is at top dead centre. This process is intended to represent the ignition of the fuel-air mixture and the subsequent rapid burning.
- Process 3–4 is an adiabatic (isentropic) expansion (power stroke).
- Process 4–1 completes the cycle by a constant-volume process in which heat is rejected from the air while the piston is at bottom dead centre.
- Process 1–0 the mass of air is released to the atmosphere in a constant pressure process.

The Otto cycle consists of isentropic compression, heat addition at constant volume, isentropic expansion, and rejection of heat at constant volume. In the case of a four-stroke Otto cycle, technically there are two additional processes: one for the exhaust of waste heat and combustion products at constant pressure (isobaric), and one for the intake of cool oxygen-rich air also at constant pressure; however, these are often omitted in a simplified analysis. Even though those two processes are critical to the functioning of a real engine, wherein the details of heat transfer and combustion chemistry are relevant, for the simplified analysis of the thermodynamic cycle, it is more convenient to assume that all of the waste-heat is removed during a single volume change.

The four-stroke engine was first patented by Alphonse Beau de Rochas in 1861. Before, in about 1854–57, two Italians (Eugenio Barsanti and Felice Matteucci) invented an engine that was rumored to be very similar, but the patent was lost.

The first person to build a working four-stroke engine, a stationary engine using a coal gas-air mixture for fuel (a gas engine), was German engineer Nikolaus Otto. This is why the four-stroke principle today is commonly known as the Otto cycle and four-stroke engines using spark plugs often are called Otto engines.

Processes

The system is defined to be the mass of air that is drawn from the atmosphere into the cylinder, compressed by the piston, heated by the spark ignition of the added fuel, allowed to expand as it pushes on the piston, and finally exhausted back into the atmosphere. The mass of air is followed as its volume, pressure and temperature change during the various thermodynamic steps. As the piston is capable of moving along the cylinder, the volume of the air changes with its position in the cylinder. The compression and expansion processes induced on the gas by the movement of the piston are idealised as reversible, i.e., no useful work is lost through turbulence or friction and no heat is transferred to or from the gas during those two processes. Energy is added to the air by the combustion of fuel. Useful work is extracted by the expansion of the gas in the cylinder. After the expansion is completed in the cylinder, the remaining heat is extracted and finally the gas is exhausted to the environment. Useful mechanical work is produced during the expansion process and some of that used to compress the air mass of the next cycle. The useful mechanical work produced minus that used for the compression process is the net work gained and that can be used for propulsion or for driving other machines. Alternatively the useful work gained is the difference between the heat added and the heat removed.

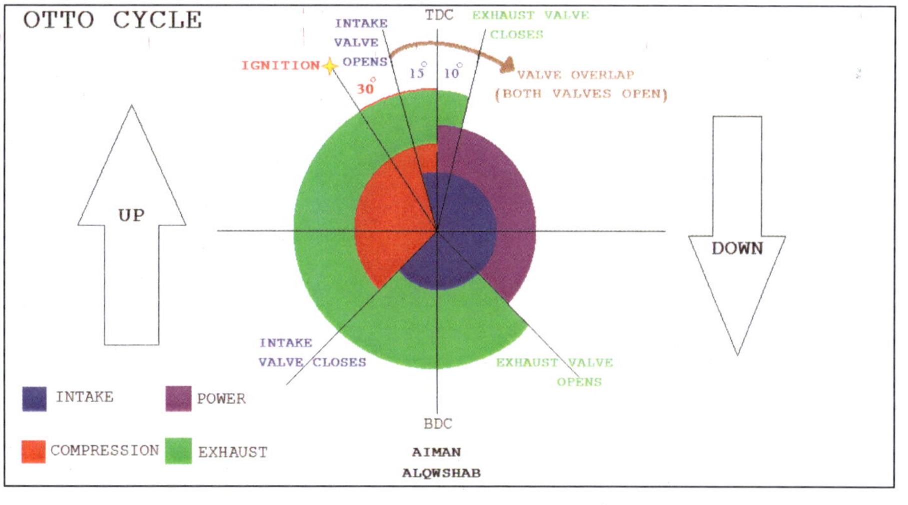

Process 0–1 Intake Stroke (Green Arrow)

A mass of air (working fluid) is drawn into the cylinder, from 0 to 1, at atmospheric pressure (constant pressure) through the open intake valve, while the exhaust valve is closed during this process. The intake valve closes at point 1.

Process 1–2 Compression Stroke (B on Diagrams)

Piston moves from crank end (BDC, bottom dead centre and maximum volume) to cylinder head end (*TDC*, top dead centre and minimum volume) as the working gas with initial state 1 is compressed isentropically to state point 2, through compression ratio (V_1/V_2). Mechanically this is the isentropic compression of the air/fuel mixture in the cylinder, also known as the compression stroke. This isentropic process assumes that no mechanical energy is lost due to friction and no heat is transferred to or from the gas, hence the process is reversible. The compression process requires that mechanical work be added to the working gas. Generally the compression ratio is around 9–10:1 (V_1:V_2) for a typical engine.

Process 2–3 Ignition Phase (C on Diagrams)

The piston is momentarily at rest at TDC. During this instant, which is known as the ignition phase, the air/fuel mixture remains in a small volume at the top of the compression stroke. Heat is added to the working fluid by the combustion of the injected fuel, with the volume essentially being held constant. The pressure rises and the ratio (P_3 / P_2) is called the "explosion ratio".

Process 3–4 Expansion Stroke (D on Diagrams)

The increased high pressure exerts a force on the piston and pushes it towards the BDC. Expansion of working fluid takes place isentropically and work is done by the system on the piston. The volume ratio V_4 / V_3 is called the "isentropic expansion ratio". (For the Otto cycle is the same as the compression ratio V_1 / V_2). Mechanically this is the expansion of the hot gaseous mixture in the cylinder known as expansion (power) stroke.

Process 4–1 Idealized Heat Rejection (A on Diagrams)

The piston is momentarily at rest at BDC. The working gas pressure drops instantaneously from point 4 to point 1 during a constant volume process as heat is removed to an idealized external sink that is brought into contact with the cylinder head. The gas has returned to state 1.

Process 1–0 Exhaust Stroke

The exhaust valve opens at point 1. As the piston moves from "BDC" (point 1) to "TDC" (point 0) with the exhaust valve opened, the gaseous mixture is vented to the atmosphere and the process starts anew.

Cycle Analysis

In the processes 1–2 the piston does work on the gas and in process 3–4 the gas does work on the piston during those isentropic compression and expansion processes, respectively. Processes 2–3 and 4–1 are isochoric processes; heat is transferred into the system from 2—3 and out of the system from 4—1 but no work is done on the system or extracted from the system during those processes. No work is done during an isochoric (constant volume) process because addition or removal of work from a system requires the movement of the boundaries of the system; hence, as the cylinder volume does not change, no shaft work is added to or removed from the system.

Four different equations are used to describe those four processes. A simplification is made by assuming changes of the kinetic and potential energy that take place in the system (mass of gas) can be neglected and then applying the first law of thermodynamics (energy conservation) to the mass of gas as it changes state as characterized by the gas's temperature, pressure, and volume.

During a complete cycle, the gas returns to its original state of temperature, pressure and volume, hence the net internal energy change of the system (gas) is zero. As a result, the energy (heat or work) added to the system must be offset by energy (heat or work) that leaves the system. In the analysis of thermodynamic systems, the convention is to account energy that enters the system as positive and energy that leaves the system is accounted as negative.

During a complete cycle, the net change of energy of the system is zero:

$$\Delta E = E_{\text{in}} - E_{\text{out}} = 0$$

The above states that the system (the mass of gas) returns to the original thermodynamic state it was in at the start of the cycle.

Where E_{in} is energy added to the system from 1–2–3 and E_{out} is energy is removed from 3–4–1. In terms of work and heat added to the system.

Equation:

$$W_{1-2} + Q_{2-3} + W_{3-4} + Q_{4-1} = 0$$

Each term of the equation can be expressed in terms of the internal energy of the gas at each point in the process:

$$W_{1-2} = U_2 - U_1$$

$$Q_{2-3} = U_3 - U_2$$

$$W_{3-4} = U_4 - U_3$$

$$Q_{4-1} = U_1 - U_4$$

The energy balance equation becomes:

$$W_{1-2} + Q_{2-3} + W_{3-4} + Q_{4-1} = (U_2 - U_1) + (U_3 - U_2) + (U_4 - U_3) + (U_1 - U_4) = 0$$

To illustrate the example we choose some values to the points in the illustration:

$$U_1 = 1$$

$$U_2 = 5$$

$$U_3 = 9$$

$$U_4 = 4$$

These values are arbitrarily but rationally selected. The work and heat terms can then be calculated.

The energy added to the system as work during the compression from 1 to 2 is:

$$(U_2 - U_1) = (5-1) = 4$$

The energy added to the system as heat from point 2 to 3 is:

$$(U_3 - U_2) = (9-5) = 4$$

The energy removed from the system as work during the expansion from 3 to 4 is:

$$(U_4 - U_3) = (4-9) = -5$$

The energy removed from the system as heat from point 4 to 1 is:

$$(U_1 - U_4) = (1-4) = -3$$

The energy balance is:

$$\Delta E = +4 + 4 - 5 - 3 = 0$$

Note that energy added to the system is counted as positive and energy leaving the system is counted as negative and the summation is zero as expected for a complete cycle that returns the system to its original state.

From the energy balance the work out of the system is:

$$\sum \text{Work} = W_{1-2} + W_{3-4} = (U_2 - U_1) + (U_4 - U_3) = 4 - 5 = -1$$

The net energy out of the system as work is -1, meaning the system has produced one net unit of energy that leaves the system in the form of work.

The net heat out of the system is:

$$\sum \text{Heat} = Q_{2-3} + Q_{4-1} = (U_3 - U_2) + (U_1 - U_4) = 4 - 3 = 1$$

As energy added to the system as heat is positive. From the above it appears as if the system gained one unit of heat. This matches the energy produced by the system as work out of the system.

Thermal efficiency is the quotient of the net work from the system, to the heat added to system. Equation:

$$\eta = \frac{W_{1-2} + W_{3-4}}{Q_{2-3}} = \frac{(U_2 - U_1) + (U_4 - U_3)}{(U_3 - U_2)}$$

$$\eta = 1 + \frac{U_1 - U_4}{(U_3 - U_2)} = 1 + \frac{(1-4)}{(9-5)} = 0.25$$

Alternatively, thermal efficiency can be derived by strictly heat added and heat rejected.

$$\eta = \frac{Q_{2-3} + Q_{4-1}}{Q_{2-3}} = 1 + \frac{(U_1 - U_4)}{(U_3 - U_2)}$$

Supplying the fictitious values

$$\eta = 1 + \frac{1-4}{9-5} = 1 + \frac{-3}{4} = 0.25$$

In the Otto cycle, there is no heat transfer during the process 1–2 and 3–4 as they are isentropic processes. Heat is supplied only during the constant volume processes 2–3 and heat is rejected only during the constant volume processes 4–1.

The above values are absolute values that might, for instance, have units of joules (assuming the MKS system of units are to be used) and would be of use for a particular engine with particular dimensions. In the study of thermodynamic systems the extensive quantities such as energy, volume, or entropy (versus intensive quantities of temperature and pressure) are placed on a unit mass basis, and so too are the calculations, making those more general and therefore of more general use. Hence, each term involving an extensive quantity could be divided by the mass, giving the terms units of joules/kg (specific energy), meters^3/kg (specific volume), or joules/(kelvin·kg) (specific entropy, heat capacity) etc. and would be represented using lower case letters, u, v, s, etc.

Equation can now be related to the specific heat equation for constant volume. The specific heats are particularly useful for thermodynamic calculations involving the ideal gas model.

$$C_v = \left(\frac{\delta u}{\delta T}\right)_v$$

Rearranging yields:

$$\delta u = (C_v)(\delta T)$$

Inserting the specific heat equation into the thermal efficiency equation yields.

$$\eta = 1 - \left(\frac{C_v(T_4 - T_1)}{C_v(T_3 - T_2)}\right)$$

Upon rearrangement:

$$\eta = 1 - \left(\frac{T_1}{T_2}\right)\left(\frac{T_4 / T_1 - 1}{T_3 / T_2 - 1}\right)$$

Next, noting from the diagrams $T_4 / T_1 = T_3 / T_2$, thus both of these can be omitted. The equation then reduces to:

Equation:

$$\eta = 1 - \left(\frac{T_1}{T_2}\right)$$

Since the Otto cycle uses isentropic processes during the compression (process 1 to 2) and expansion (process 3 to 4) the isentropic equations of ideal gases and the constant pressure/volume relations can be used to yield equations.

Equation:

$$\left(\frac{T_2}{T_1}\right)=\left(\frac{p_2}{p_1}\right)^{(\gamma-1)/\gamma}$$

Equation:

$$\left(\frac{T_2}{T_1}\right)=\left(\frac{V_1}{V_2}\right)^{(\gamma-1)}$$

Where,

$$\gamma=\left(\frac{C_{\mathrm{p}}}{C_{\mathrm{v}}}\right)$$

γ is the specific heat ratio.

The derivation of the previous equations are found by solving these four equations respectively (where R is the specific gas constant):

$$C_{\mathrm{p}}\ln\left(\frac{V_1}{V_2}\right)-R\ln\left(\frac{p_2}{p_1}\right)=0$$

$$C_v\ln(\frac{T_2}{T_1})-R\ln(\frac{V_2}{V_1})=0$$

$$C_{\mathrm{p}}=\left(\frac{\gamma R}{\gamma-1}\right)$$

$$C_{\mathrm{v}}=\left(\frac{R}{\gamma-1}\right)$$

Further simplifying equation, where r is the compression ratio (V_1/V_2):

Equation:

$$\left(\frac{T_2}{T_1}\right)=\left(\frac{V_1}{V_2}\right)^{(\gamma-1)}=r^{(\gamma-1)}$$

From inverting equation and inserting it into equation the final thermal efficiency can be expressed as: namec = "Fundamentals".

Equation:

$$\eta = 1 - \left(\frac{1}{r^{(\gamma-1)}} \right)$$

From analyzing equation it is evident that the Otto cycle efficiency depends directly upon the compression ratio r. Since the γ for air is 1.4, an increase in r will produce an increase in η. However, the γ for combustion products of the fuel/air mixture is often taken at approximately 1.3. The foregoing discussion implies that it is more efficient to have a high compression ratio. The standard ratio is approximately 10:1 for typical automobiles. Usually this does not increase much because of the possibility of autoignition, or "knock", which places an upper limit on the compression ratio. During the compression process 1–2 the temperature rises, therefore an increase in the compression ratio causes an increase in temperature. Autoignition occurs when the temperature of the fuel/air mixture becomes too high before it is ignited by the flame front. The compression stroke is intended to compress the products before the flame ignites the mixture. If the compression ratio is increased, the mixture may auto-ignite before the compression stroke is complete, leading to "engine knocking". This can damage engine components and will decrease the brake horsepower of the engine.

Power

The power produced by the Otto cycle is the energy developed per unit of time. The Otto engines are called four-stroke engines. The intake stroke and compression stroke require one rotation of the engine crankshaft. The power stroke and exhaust stroke require another rotation. For two rotations there is one work generating stroke.

From the above cycle analysis the net work produced by the system:

$$\sum \text{Work} = W_{1-2} + W_{3-4} = (U_2 - U_1) + (U_4 - U_3) = +4 - 5 = -1$$

(again, using the sign convention, the minus sign implies energy is leaving the system as work)

If the units used were MKS the cycle would have produced one joule of energy in the form of work. For an engine of a particular displacement, such as one liter, the mass of gas of the system can be calculated assuming the engine is operating at standard temperature (20 °C) and pressure (1 atm). Using the Universal Gas Law the mass of one liter of gas is at room temperature and sea level pressure:

$$M = \frac{PV}{RT}$$

V=0.001 m^3, R=0.286 kJ/(kg·K), T=293 K, P=101.3 kN/m^2

M=0.00121 kg

At an engine speed of 3000 RPM there are 1500 work-strokes/minute or 25 work-strokes/second.

$$\sum \text{Work} = 1\text{J} / (\text{kg·stroke}) \times 0.00121\text{kg} = 0.00121\text{J} / \text{stroke}$$

Power is 25 times that since there are 25 work-strokes/second.

$$P = 25 \times 0.00121 = 0.0303\text{J}/\text{s or W}$$

If the engine is multi-cylinder, the result would be multiplied by that factor. If each cylinder is of a different liter displacement, the results would also be multiplied by that factor. These results are the product of the values of the internal energy that were assumed for the four states of the system at the end each of the four strokes (two rotations). They were selected only for the sake of illustration, and are obviously of low value. Substitution of actual values from an actual engine would produce results closer to that of the engine. Whose results would be higher than the actual engine as there are many simplifying assumptions made in the analysis that overlook inefficiencies. Such results would overestimate the power output.

Increasing Power and Efficiency

The difference between the exhaust and intake pressures and temperatures means that some increase in efficiency can be gained by use of a turbocharger, removing from the exhaust flow some part of the remaining energy and transferring that to the intake flow to increase the intake pressure. A gas turbine can extract useful work energy from the exhaust stream and that can then be used to pressurize the intake air. The pressure and temperature of the exhausting gases would be reduced as they expand through the gas turbine and that work is then applied to the intake gas stream, increasing its pressure and temperature. The transfer of energy amounts to an efficiency improvement and the resulting power density of the engine is also improved. The intake air is typically cooled so as to reduce its volume as the work produced per stroke is a direct function of the amount of mass taken into the cylinder; denser air will produce more work per cycle. Practically speaking the intake air mass temperature must also be reduced to prevent premature ignition in a petrol fueled engine; hence, an intercooler is used to remove some energy as heat and so reduce the intake temperature. Such a scheme both increases the engine's efficiency and power.

The application of a supercharger driven by the crankshaft does increase the power output (power density) but does not increase efficiency as it uses some of the net work produced by the engine to pressurize the intake air and fails to extract otherwise wasted energy associated with the flow of exhaust at high temperature and a pressure to the ambient.

CARNOT CYCLE

The Carnot cycle is a theoretical ideal thermodynamic cycle proposed by French physicist Sadi Carnot in 1824 and expanded upon by others in the 1830s and 1840s. It provides an upper limit on the efficiency that any classical thermodynamic engine can achieve during the conversion of heat into work, or conversely, the efficiency of a refrigeration system in creating a temperature difference by the application of work to the system. It is not an actual thermodynamic cycle but is a theoretical construct.

Every single thermodynamic system exists in a particular state. When a system is taken through a series of different states and finally returned to its initial state, a thermodynamic cycle is said to have occurred. In the process of going through this cycle, the system may perform work on its surroundings, for example by moving a piston, thereby acting as a heat engine. A system undergoing a Carnot cycle is called a Carnot heat engine, although such a "perfect" engine is only a theoretical construct and cannot be built in practice. However, a microscopic Carnot heat engine has been designed and run.

Essentially, there are two "heat reservoirs" forming part of the heat engine at temperatures T_h and T_c (hot and cold respectively). They have such large thermal capacity that their temperatures are practically unaffected by a single cycle. Since the cycle is theoretically reversible, there is no generation of entropy during the cycle; entropy is conserved. During the cycle, an arbitrary amount of entropy ΔS is extracted from the hot reservoir, and deposited in the cold reservoir. Since there is no volume change in either reservoir, they do no work, and during the cycle, an amount of energy $T_h\Delta S$ is extracted from the hot reservoir and a smaller amount of energy $T_c\Delta S$ is deposited in the cold reservoir. The difference in the two energies $(T_h-T_c)\Delta S$ is equal to the work done by the engine.

Stages

The Carnot cycle when acting as a heat engine consists of the following steps:

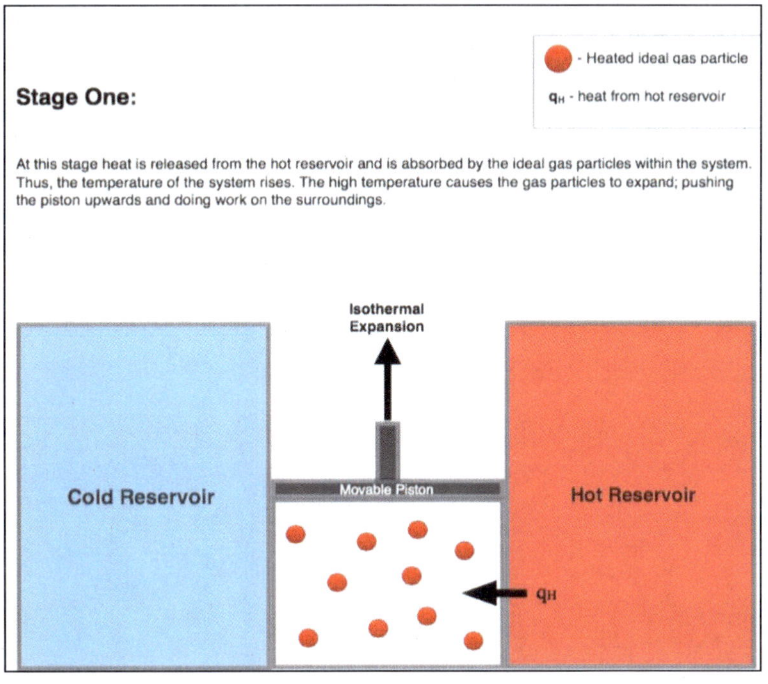

- Isothermal Expansion: Heat is transferred reversibly from high temperature reservoir at constant temperature T_H (isothermal heat addition or absorption). During this step (1 to 2 on figure , A to B in figure) the gas is allowed to expand, doing work on the surroundings by pushing up the piston (stage 1 figure, right). Although the pressure drops from points 1 to 2 the temperature of the gas does not change during the process because it is in thermal contact with the hot reservoir at T_h, and thus the expansion is isothermal. Heat energy Q_1 is absorbed from the high temperature reservoir resulting in an increase in the entropy of the gas by the amount $\Delta S_1 = Q_1 / T_h$.

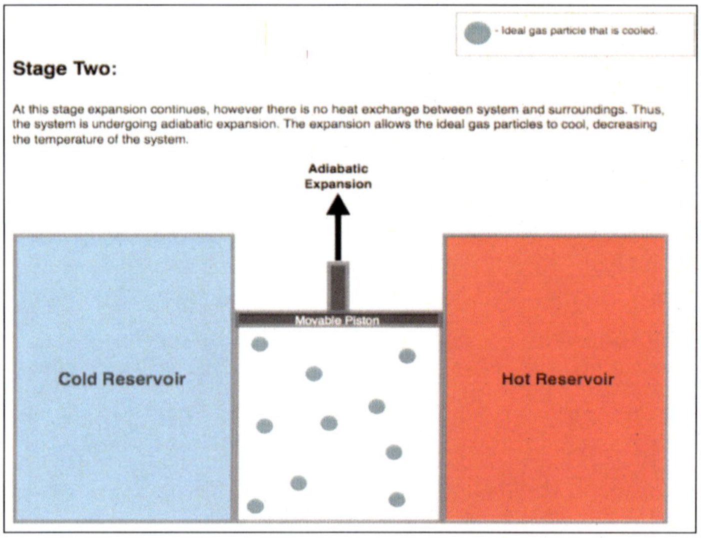

- Isentropic (reversible adiabatic) expansion of the gas (isentropic work output): For this step (2 to 3 on figure, B to C in figure) the gas in the engine is thermally insulated from both the hot and cold reservoirs. Thus they neither gain nor lose heat, an 'adiabatic' process. The gas continues to expand by reduction of pressure, doing work on the surroundings (raising the piston; stage 2 figure, right), and losing an amount of internal energy equal to the work done. The gas expansion without heat input causes it to cool to the "cold" temperature, T_c. The entropy remains unchanged.

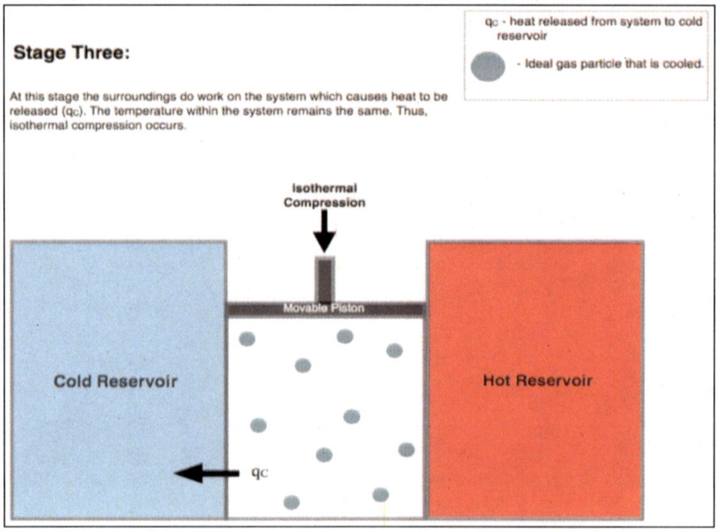

- Isothermal Compression: Heat transferred reversibly to low temperature reservoir at constant temperature T_C. (isothermal heat rejection) (3 to 4 on figure, C to D on figure) Now the gas in the engine is in thermal contact with the cold reservoir at temperature T_c. The surroundings do work on the gas, pushing the piston down (stage 3 figure, right), causing an amount of heat energy Q_2 to leave the system to the low temperature reservoir and the entropy of the system to decrease by the amount $\Delta S_2 = Q_2 / T_c$. (This is the same amount of entropy absorbed in step 1, as can be seen from the Clausius inequality.)

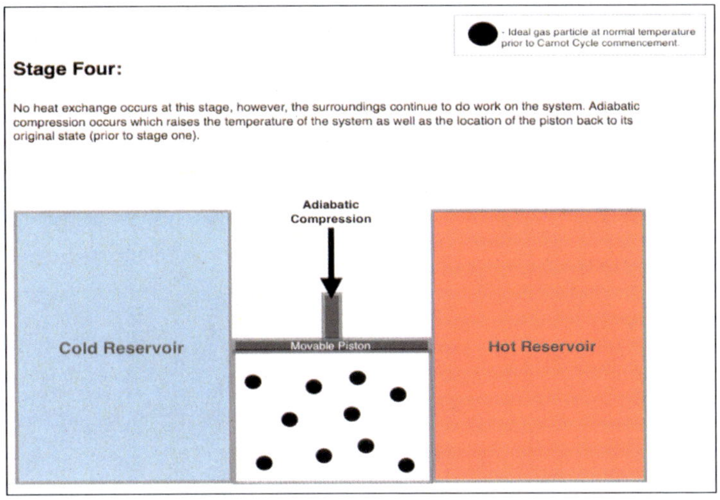

- Adiabatic reversible compression: (4 to 1 on figure , D to A on figure) Once again the gas in the engine is thermally insulated from the hot and cold reservoirs, and the engine is assumed to be frictionless, hence reversible. During this step, the surroundings do work on the gas, pushing the piston down further (stage 4 figure, right), increasing its internal energy, compressing it, and causing its temperature to rise back to Th due solely to the work added to the system, but the entropy remains unchanged. At this point the gas is in the same state as at the start of step 1.

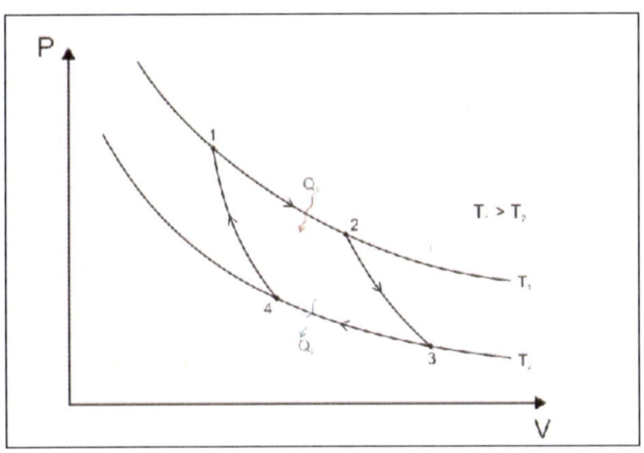

A Carnot cycle illustrated on a PV diagram to illustrate the work done.

In this case,

$$\Delta S_1 = \Delta S_2,$$

or,

$$Q_1 / T_h = Q_2 / T_c.$$

This is true as Q_2 and T_c are both lower and in fact are in the same ratio as Q_1 / T_h.

The Pressure–volume Graph

When the Carnot cycle is plotted on a pressure–volume diagram, the isothermal stages follow the isotherm lines for the working fluid, the adiabatic stages move between isotherms, and the area bounded by the complete cycle path represents the total work that can be done during one cycle. From point 1 to 2 and point 3 to 4 the temperature is constant. Heat transfer from point 4 to 1 and point 2 to 3 are equal to zero.

Properties and Significance

The Temperature–entropy Diagram

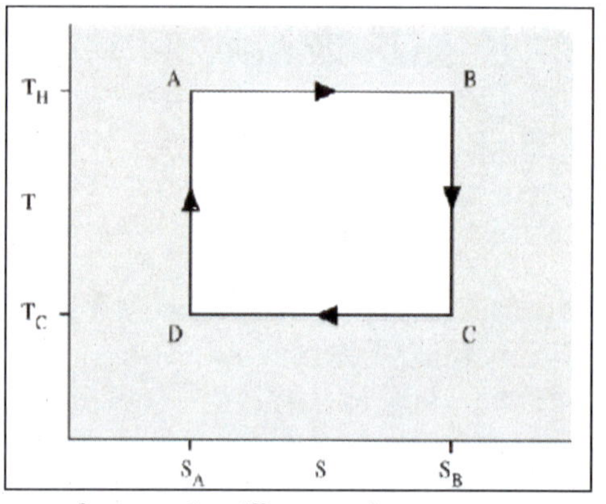

A Carnot cycle acting as a heat engine, illustrated on a temperature–entropy diagram. The cycle takes place between a hot reservoir at temperature TH and a cold reservoir at temperature TC. The vertical axis is temperature, the horizontal axis is entropy.

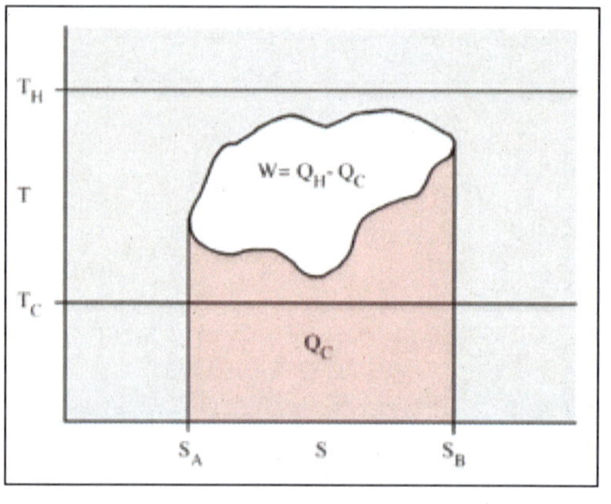

A generalized thermodynamic cycle taking place between a hot reservoir at temperature T_H and a cold reservoir at temperature T_C. By the second law of thermodynamics, the cycle cannot extend outside the temperature band from T_C to T_H. The area in red Q_C is the amount of energy exchanged between the system and the cold reservoir. The area in white W is the amount of work energy exchanged by the system with its surroundings. The amount of heat exchanged with the hot reservoir is the sum of the two. If the system is behaving as an engine, the process moves clockwise around the loop, and moves counter-clockwise if it is behaving as a refrigerator. The efficiency to the cycle is the ratio of the white area (work) divided by the sum of the white and red areas (heat absorbed from the hot reservoir).

The behaviour of a Carnot engine or refrigerator is best understood by using a temperature–entropy diagram (T–S diagram), in which the thermodynamic state is specified by a point on a graph with entropy (S) as the horizontal axis and temperature (T) as the vertical axis. For a simple closed system (control mass analysis), any point on the graph will represent a particular state of the system. A thermodynamic process will consist of a curve connecting an initial state (A) and a final state (B). The area under the curve will be:

$$Q = \int_A^B T\,dS$$

which is the amount of thermal energy transferred in the process. If the process moves to greater entropy, the area under the curve will be the amount of heat absorbed by the system in that process. If the process moves towards lesser entropy, it will be the amount of heat removed. For any cyclic process, there will be an upper portion of the cycle and a lower portion. For a clockwise cycle, the area under the upper portion will be the thermal energy absorbed during the cycle, while the area under the lower portion will be the thermal energy removed during the cycle. The area inside the cycle will then be the difference between the two, but since the internal energy of the system must have returned to its initial value, this difference must be the amount of work done by the system over the cycle. Referring to figure, mathematically, for a reversible process we may write the amount of work done over a cyclic process as:

$$W = \oint PdV = \oint (dQ - dU) = \oint (TdS - dU) = \oint TdS - \oint dU = \oint TdS$$

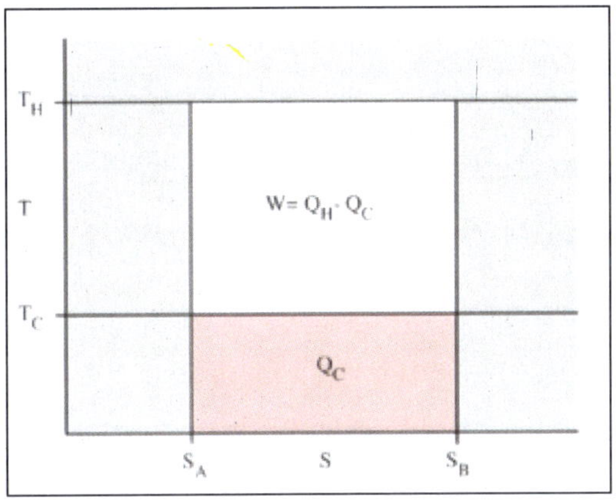

A Carnot cycle taking place between a hot reservoir at temperature T_H and a cold reservoir at temperature T_C.

Since dU is an exact differential, its integral over any closed loop is zero and it follows that the area inside the loop on a T–S diagram is equal to the total work performed if the loop is traversed in a clockwise direction, and is equal to the total work done on the system as the loop is traversed in a counterclockwise direction.

The Carnot Cycle

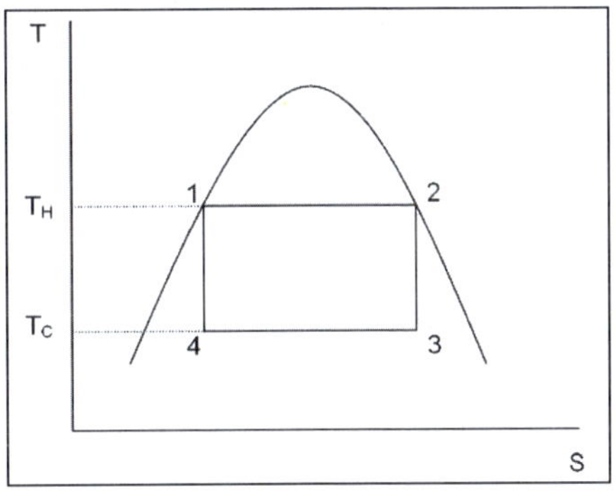

A visualization of the Carnot cycle.

Evaluation of the above integral is particularly simple for the Carnot cycle. The amount of energy transferred as work is:

$$W = \oint PdV = \oint TdS = (T_H - T_C)(S_B - S_A)$$

The total amount of thermal energy transferred from the hot reservoir to the system will be:

$$Q_H = T_H(S_B - S_A)$$

and the total amount of thermal energy transferred from the system to the cold reservoir will be:

$$Q_C = T_C(S_B - S_A)$$

The efficiency η is defined to be:

$$\eta = \frac{W}{Q_H} = \frac{Q_H - Q_C}{Q_H} = 1 - \frac{T_C}{T_H}$$

where

W is the work done by the system (energy exiting the system as work),

Q_C is the heat taken from the system (heat energy leaving the system),

Q_H is the heat put into the system (heat energy entering the system),

T_C is the absolute temperature of the cold reservoir,

T_H is the absolute temperature of the hot reservoir,

S_B is the maximum system entropy,

S_A is the minimum system entropy.

This definition of efficiency makes sense for a heat engine, since it is the fraction of the heat energy extracted from the hot reservoir and converted to mechanical work. A Rankine cycle is usually the practical approximation.

Reversed Carnot Cycle

The Carnot heat-engine cycle described is a totally reversible cycle. That is all the processes that compose it can be reversed, in which case it becomes the Carnot refrigeration cycle. This time, the cycle remains exactly the same except that the directions of any heat and work interactions are reversed. Heat is absorbed from the low-temperature reservoir, heat is rejected to a high-temperature reservoir, and a work input is required to accomplish all this. The P–V diagram of the reversed Carnot cycle is the same as for the Carnot cycle except that the directions of the processes are reversed.

Carnot's Theorem

It can be seen from the above diagram, that for any cycle operating between temperatures T_H and T_C, none can exceed the efficiency of a Carnot cycle.

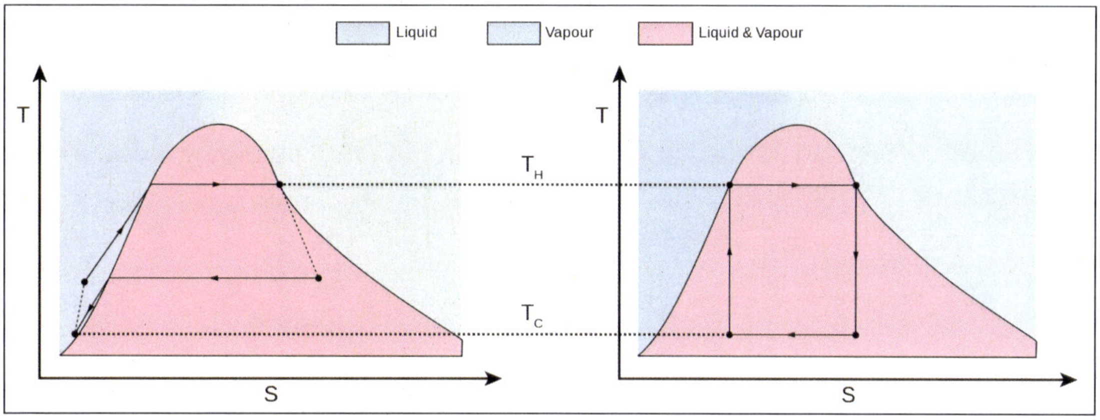

A real engine (left) compared to the Carnot cycle (right). The entropy of a real material changes with temperature. This change is indicated by the curve on a T–S diagram. For this figure, the curve indicates a vapor-liquid equilibrium. Irreversible systems and losses of energy (for example, work due to friction and heat losses) prevent the ideal from taking place at every step.

Carnot's theorem is a formal statement of this fact: No engine operating between two heat reservoirs can be more efficient than a Carnot engine operating between those same reservoirs. Thus, equation gives the maximum efficiency possible for any engine using the corresponding temperatures. A corollary to Carnot's theorem states that. All reversible engines operating between the same heat reservoirs are equally efficient. Rearranging the right side of the equation gives what may be a more easily understood form of the equation, namely that the theoretical maximum efficiency of a heat engine equals the difference in temperature between the hot and cold reservoir divided by the absolute temperature of the hot reservoir. Looking at this formula an interesting fact becomes apparent. Lowering the temperature of the cold reservoir will have more effect on the ceiling efficiency of a heat engine than raising the temperature of the hot reservoir by the same amount. In the real world, this may be difficult to achieve since the cold reservoir is often an existing ambient temperature.

In other words, maximum efficiency is achieved if and only if no new entropy is created in the cycle, which would be the case if e.g. friction leads to dissipation of work into heat. In that case, the cycle is not reversible and the Clausius theorem becomes an inequality rather than an equality. Otherwise, since entropy is a state function, the required dumping of heat into the environment to dispose of excess entropy leads to a (minimal) reduction in efficiency. So equation gives the efficiency of any reversible heat engine.

In mesoscopic heat engines, work per cycle of operation in general fluctuates due to thermal noise. If the cycle is performed quasi-statically, the fluctuations vanish even on the mesoscale. However, if the cycle is performed faster than the relaxation time of the working medium, the fluctuations of work are inevitable. Nevertheless, when work and heat fluctuations are counted, there is exact equality that relates the exponential average of work performed by any heat engine and the heat transfer from the hotter heat bath.

Efficiency of Real Heat Engines

Carnot realized that in reality it is not possible to build a thermodynamically reversible engine, so real heat engines are even less efficient than indicated by equation. In addition, real engines that operate along this cycle are rare. Nevertheless, equation is extremely useful for determining the maximum efficiency that could ever be expected for a given set of thermal reservoirs.

Although Carnot's cycle is an idealisation, the expression of Carnot efficiency is still useful. Consider the average temperatures,

$$\langle T_H \rangle = \frac{1}{\Delta S} \int_{Q_{in}} T dS$$

$$\langle T_C \rangle = \frac{1}{\Delta S} \int_{Q_{out}} T dS$$

at which heat is input and output, respectively. Replace T_H and T_C in equation by $(T_H$ and (T_C) respectively.

For the Carnot cycle, or its equivalent, the average value (T_H) will equal the highest temperature available, namely T_H, and (T_C) the lowest, namely T_C. For other less efficient cycles, (T_H) will be lower than T_H, and (T_C) will be higher than T_C. This can help illustrate, for example, why a reheater or a regenerator can improve the thermal efficiency of steam power plants—and why the thermal efficiency of combined-cycle power plants (which incorporate gas turbines operating at even higher temperatures) exceeds that of conventional steam plants. The first prototype of the diesel engine was based on the Carnot cycle.

DIESEL CYCLE

Diesel cycle is one of most common thermodynamic cycles that can be found in automobile engines and describes the functioning of a typical compression ignition piston engine. The Diesel engine is similar in operation to the gasoline engine. The most important difference is that:

- There is no fuel in the cylinder at the beginning of the compression stroke, therefore an autoignition does not occur in Diesel engines.
- Diesel engine uses compression ignition instead of spark ignition.
- Because of the high temperature developed during the adiabatic compression, the fuel ignites spontaneously as it is injected. Therefore no spark plugs are needed.
- Before the beginning of the power stroke, the injectors start to inject fuel directly into the combustion chamber and therefore first part of power stroke occurs approximately at the constant pressure.
- Higher compression ratios can be achieved in Diesel engines, than in Otto engines.

The Diesel engine is similar in operation to the gasoline engine. In this picture, there is an Otto engine, which is ignited by a spark plug instead of compression itself.

In contrast to Otto cycle, the Diesel cycle does not execute isochoric heat addition. In an ideal Diesel cycle, the system executing the cycle undergoes a series of four processes: two isentropic (reversible adiabatic) processes alternated with one isochoric process and one isobaric process.

Since Carnot's principle states that no engine can be more efficient than a reversible engine (a Carnot heat engine) operating between the same high temperature and low temperature reservoirs, the Diesel engine must have lower efficiency than the Carnot efficiency. A typical diesel automotive engine operates at around 30% to 35% of thermal efficiency. About 65-70% is rejected as waste heat without being converted into useful work, i.e. work delivered to wheels. In general, engines using the Diesel cycle are usually more efficient, than engines using the Otto cycle. The diesel engine has the highest thermal efficiency of any practical combustion engine. Low-speed diesel engines (as used in ships) can have a thermal efficiency that exceeds 50%. The largest diesel engine in the world peaks at 51.7%.

Processes

In an ideal Diesel cycle, the system executing the cycle undergoes a series of four processes: two isentropic (reversible adiabatic) processes alternated with one isochoric process and one isobaric process:

- Isentropic compression (compression stroke) – The air is compressed adiabatically from state 1 to state 2, as the piston moves from bottom dead center to top dead center. The surroundings do work on the gas, increasing its internal energy (temperature) and compressing it. On the other hand the entropy remains unchanged. The changes in volumes and its ratio (V1 / V2) is known as the compression ratio.
- Isobaric expansion (ignition phase) – In this phase (between state 2 and state 3) there is a constant-pressure (idealized model) heat transfer to the air from an external source (combustion of injected fuel) while the piston is moving toward the V3. During the constant pressure process, energy enters the system as heat Qadd, and a part of work is done by moving piston.
- Isentropic expansion (power stroke) – The gas expands adiabatically from state 3 to state 4, as the piston moves from V3 to bottom dead center. The gas does work on the surroundings (piston) and loses an amount of internal energy equal to the work that leaves the system. Again the entropy remains unchanged. The volume ratio (V4 / V3) is known as the isentropic expansion ratio.
- Isochoric decompression (exhaust stroke) – In this phase the cycle completes by a constant-volume process in which heat is rejected from the air while the piston is at bottom dead center. The working gas pressure drops instantaneously from point 4 to point 1. The exhaust valve opens at point 4. The exhaust stroke is directly after this decompression. As the piston moves from bottom dead center (point 1) to top dead center (point 0) with the exhaust valve opened, the gaseous mixture is vented to the atmosphere and the process starts anew.

During the Diesel cycle, work is done on the gas by the piston between states 1 and 2 (isentropic compression). Work is done by the gas on the piston between stages 2 and 3 (isobaric heat addition) and between stages 2 and 3 (isentropic expansion). The difference between the work done by the gas and the work done on the gas is the net work produced by the cycle and it corresponds to the area enclosed by the cycle curve. The work produced by the cycle times the rate of the cycle (cycles per second) is equal to the power produced by the Diesel engine.

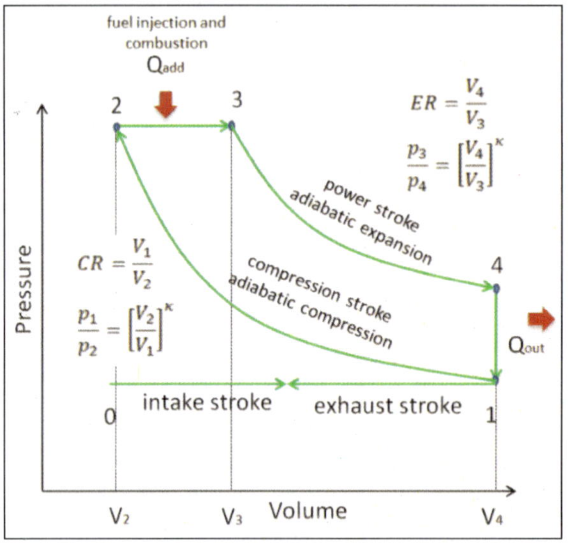

pV diagram of an ideal Diesel cycle.

Isentropic Process

An isentropic process is a thermodynamic process, in which the entropy of the fluid or gas remains constant. It means the isentropic process is a special case of an adiabatic process in which there is no transfer of heat or matter. It is a reversible adiabatic process. The assumption of no heat transfer is very important, since we can use the adiabatic approximation only in very rapid processes.

Isentropic Process and the First Law

For a closed system, we can write the first law of thermodynamics in terms of enthalpy:

$$dH = dQ + Vdp$$

or

$$dH = TdS + Vdp$$

Isentropic process (dQ = 0):

$$dH = Vdp \rightarrow W = H2 - H1 \rightarrow H2 - H1 = Cp\,(T2 - T1) \text{ (for ideal gas)}$$

Isentropic Process of the Ideal Gas

The isentropic process (a special case of adiabatic process) can be expressed with the ideal gas law as:

$$pV^{\kappa} = \textit{constant}$$

or

$$p_1V_1^{\kappa} = p_2V_2^{\kappa}$$

in which κ = cp/cv is the ratio of the specific heats (or heat capacities) for the gas. One for constant pressure (cp) and one for constant volume (cv). Note that, this ratio κ = cp/cv is a factor in determining the speed of sound in a gas and other adiabatic processes.

Isochoric Process

An isochoric process is a thermodynamic process, in which the volume of the closed system remains constant (V = const). It describes the behavior of gas inside the container, that cannot be deformed. Since the volume remains constant, the heat transfer into or out of the system does not the pΔV work, but only changes the internal energy (the temperature) of the system.

Isochoric Process and the First Law

The classical form of the first law of thermodynamics is the following equation:

$$dU = dQ - dW$$

In this equation dW is equal to dW = pdV and is known as the boundary work. Then:

$$dU = dQ - pdV$$

In isochoric process and the ideal gas, all of heat added to the system will be used to increase the internal energy.

Isochoric process (pdV = 0):

$$dU = dQ \text{ (for ideal gas)}$$

$$dU = 0 = Q - W \rightarrow W = Q \text{ (for ideal gas)}$$

Isochoric Process of the Ideal Gas

The isochoric process can be expressed with the ideal gas law as:

$$\frac{p}{T} = \text{constant}$$

Or

$$\frac{p_1}{T_1} = \frac{p_2}{T_2}$$

On a p-V diagram, the process occurs along a horizontal line that has the equation V = constant.

Isobaric Process

An isobaric process is a thermodynamic process, in which the pressure of the system remains constant (p = const). The heat transfer into or out of the system does work, but also changes the internal energy of the system.

Since there are changes in internal energy (dU) and changes in system volume (ΔV), engineers often use the enthalpy of the system, which is defined as:

$$H = U + pV$$

Isobaric Process and the First Law

The classical form of the first law of thermodynamics is the following equation:

$$dU = dQ - dW$$

In this equation dW is equal to dW = pdV and is known as the boundary work. In an isobaric process and the ideal gas, part of heat added to the system will be used to do work and part of heat added will increase the internal energy (increase the temperature). Therefore it is convenient to use the enthalpy instead of the internal energy.

Isobaric process (Vdp = 0):

$$dH = dQ \rightarrow Q = H_2 - H_1$$

At constant entropy, i.e. in isentropic process, the enthalpy change equals the flow process work done on or by the system.

Isobaric Process of the Ideal Gas

The isobaric process can be expressed with the ideal gas law as:

$$\frac{V}{T} = \text{Constant}$$

Or

$$\frac{V_1}{T_1} = \frac{V_2}{T_2}$$

On a p-V diagram, the process occurs along a horizontal line (called an isobar) that has the equation p = constant.

Isobaric process – main characteristics

First Law	$dQ = TdS = 0$ $dU = pdV$ $dH = Vdp$
Ideal Gas Relation	$pV^k = \text{constant}$
P,V, T Relations	$\frac{p_1}{p_2} = \left[\frac{V_2}{V_1}\right]^k = \left[\frac{T_1}{T_2}\right]^{\frac{k}{k-1}}$

Change in Internal Energy	$dU = mc_V(T_2 - T_1)$
Change in Enthalpy	$dH = mc_P(T_2 - T_1)$
Heat Transfer	O
pdV Work	$W_{dv} = pdV = dU = mc_V(T_2 - T_1)$
Vdp Work	$W_{dp} = Vdp = dH = mc_P(T_2 - T_1)$
First Law	$dU = dQ$
Ideal Gas Relation	$\frac{p}{T} = \text{constant}$
P,V, T Relations	$\frac{p_1}{T_1} = \frac{p_2}{T_2}$
Change in Internal Energy	$dU = mc_V(T_2 - T_1)$

Heat Transfer	$dQ = mc_V(T_2 - T_1)$
pdV Work	$W_{i \to f} = 0$
First Law	$dU = dQ - pdV$ $dH = dQ - Vdp = dQ$
Ideal Gas Relation	$\frac{V}{T} = \text{constant}$
P,V, T Relations	$\frac{V_1}{T_1} = \frac{V_2}{T_2}$
Change in Internal Energy	$dU = mc_V(T_2 - T_1)$
Change in Enthalpy	$dH = mc_P(T_2 - T_1)$
Heat Transfer	$dQ = mc_P(T_2 - T_1)$
pdV Work	$W_{i \to f} = p(V_f - V_i)$ $W_{i \to f} = nR(T_f - T_i)$
Vdp Work	$W_{i \to f} = 0$

Diesel Cycle – pV, Ts Diagram

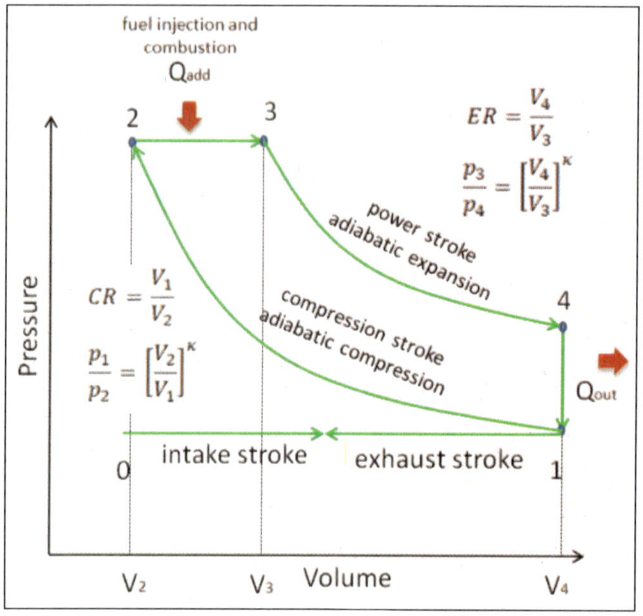

pV diagram of an ideal Diesel cycle.

Diesel cycles are often plotted on a pressure-volume diagram (pV diagram) and on a temperature-entropy diagram (Ts diagram).

When plotted on a pressure volume diagram, the isobaric process follow the isobar line for the gas (the horizontal lines), the isochoric process follow the isochoric line for the gas (the vertical line), adiabatic processes move between these lines and the area bounded by the complete cycle path represents the total work that can be done during one cycle.

The temperature-entropy diagram (Ts diagram) in which the thermodynamic state is specified by a point on a graph with specific entropy (s) as the horizontal axis and absolute temperature (T) as the vertical axis. Ts diagrams are a useful and common tool, particularly because it helps to visualize the heat transfer during a process. For reversible (ideal) processes, the area under the T-s curve of a process is the heat transferred to the system during that process.

Four Stroke Diesel Engine

Diesel engines may be designed as either two stroke or four stroke cycles.The four stroke Diesel engine is an internal combustion (IC) engine in which the piston completes four separate strokes while turning a crankshaft. A stroke refers to the full travel of the piston along the cylinder, in either direction.

The Four Stroke Engine Comprises

- The intake stroke – The piston moves from top dead center (TDC) to bottom dead center (BDC) and the cycle passes points $0 \rightarrow 1$. In this stroke the intake valve is open while the piston pulls air (without a fuel) into the cylinder by producing vacuum pressure into the cylinder through its downward motion.

- The compression stroke – The piston moves from bottom dead center (BDC) to top dead center (TDC) and the cycle passes points 1 → 2. In this stroke both the intake and exhaust valves are closed, resulting in adiabatic air compression (i.e. without heat transfer to or from the environment). During this compression, the volume is reduced, the pressure and temperature both rise. At the end of this stroke fuel is injected and burns in the compressed hot air. At the end of this stroke the crankshaft has completed a full 360 degree revolution.
- The power stroke – The piston moves from top dead center (TDC) to bottom dead center (BDC) and the cycle passes points 2 → 3 → 4. In this stroke both the intake and exhaust valves are closed. At the beginning of the power stroke, a near isobaric combustion occur between 2 and 3. In this interval the pressure remains constant since the piston descends, and the volume increases. At 3 fuel injection and combustion are complete, and the cylinder contains gas at a higher temperature than at 2. Between 3 and 4 this hot gas expands, again approximately adiabatically. In this stroke the piston is driven towards the crankshaft, the volume in increased, and the work is done by the gas on the piston.
- The exhaust stroke. The piston moves from bottom dead center (BDC) to top dead center (TDC) and the cycle passes points 4 → 1 → 0. In this stroke the exhaust valve is open while the piston pulls an exhaust gases out of the chamber. At the end of this stroke the crankshaft has completed a second full 360 degree revolution.

In an ideal case, the adiabatic expansion should continue, until the pressure falls to that of the surrounding air. This would increase the thermal efficiency of such engine, but this also causes the practical difficulties with the engine. Simply the engine would have to be much larger.

Comparison of Actual and Ideal Diesel Cycles

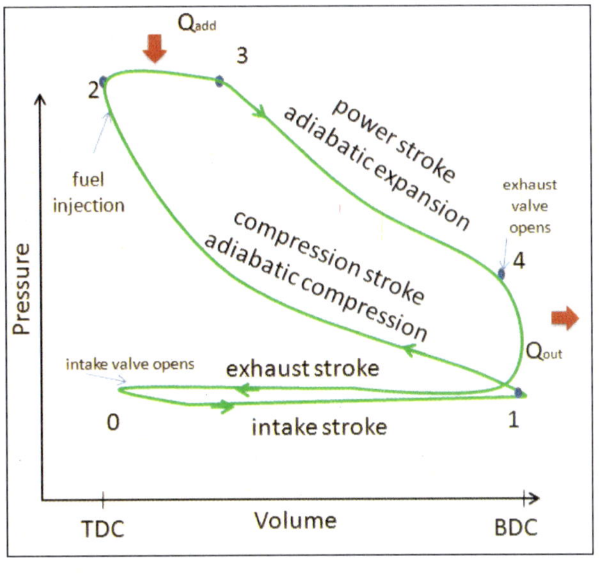

An ideal Diesel cycle in which there are a lot of assumptions differs from actual Diesel cycle. The main differences between the actual and ideal Diesel engine appear in the figure. In reality, the ideal cycle does not occur and there are many losses associated with each process. For an actual cycle, the shape of the pV diagram is similar to the ideal, but the area (work) enclosed by the pV diagram is always less than the ideal value. The ideal Diesel cycle is based on the following assumptions:

- Closed cycle: The largest difference between the two diagrams is the simplification of the intake and exhaust strokes in the ideal cycle. In the exhaust stroke, heat Qout is ejected to the environment (in a real engine, the gas leaves the engine and is replaced by a new mixture of air and fuel).
- Isobaric heat addition: In real engines the heat addition is never isobaric.
- No heat transfer:
    - Compression: The gas is compressed adiabatically from state 1 to state 2. In real engines, there are always some inefficiencies that reduce the thermal efficiency.
    - Expansion: The gas expands adiabatically from state 3 to state 4.
- Complete combustion of the mixture.
- No pumping work: Pumping work is the difference between the work done during exhaust stroke and the work done during intake stroke. In real cycles, there is a pressure difference between exhaust and inlet pressures.
- No blowdown loss: Blowdown loss is caused by the early opening of exhaust valves. This results in a loss of work output during expansion stroke.
- No blow-by loss: The blow-by loss is caused by the leakage of compressed gases through piston rings and other crevices.
- No frictional losses.

These simplifying assumptions and losses lead to the fact that the enclosed area (work) of the pV diagram for an actual engine is significantly smaller than the size of the area (work) enclosed by the pV diagram of the ideal cycle. In other words, the ideal engine cycle will overestimate the net work and, if the engines run at the same speed, greater power produced by the actual engine by around 20% (similarly as in the case of Otto engine).

Compression Ratio – Otto Engine

The compression ratio, CR, is defined as the ratio of the volume at bottom dead center and the volume at top dead center. It is a key characteristics for many internal combustion engines. the compression ratio determines the thermal efficiency of used thermodynamic cycle of the combustion engine. In general, it is desired to have a high compression ratio, because it allows an engine to reach higher thermal efficiency.

For example, let assume an Otto cycle with compression ratio of CR = 10 : 1. The volume of the chamber is 500 cm^3 = 500×10-6 m3 (0.5l) prior to the compression stroke. For this engine all required volumes are known:

- $V_1 = V_4 = V_{max} = 500\times10^{-6}\ m^3$ (0.5l)

- $V_2 = V_3 = V_{min} = V_{max} / CR = 55.56 \times 10^{-6}\ m^3$

Note that ($V_{max} - V_{min}$) x number of cylinders = total engine displacement.

Examples of Compression Ratios – Gasoline vs. Diesel

- The compression ratio in a gasoline-powered engine will usually not be much higher than 10:1 due to potential engine knocking (autoignition) and not lower than 6:1.
- A turbocharged Subaru Impreza WRX has a compression ratio of 8.0:1. In general, a turbocharged or supercharged engines already have compressed air at air intake, therefore they are usually built with lower compression ratio.
- A stock Honda S2000 engine (F22C1) has a compression ratio of 11.1:1.
- Some atmospheric sportscar engines can have compression ratio up to 12.5 : 1 (e.g. Ferrari 458 Italia).
- In 2012, Mazda released new petrol engines under the brand name SkyActiv with a 14:1 compression ratio. To reduce the risk of engine knocking, residual gas is reduced by using 4-2-1 engine exhaust systems, implementing a piston cavity, and optimizing fuel injection.
- The Diesel engines have the compression ratio that normally exceed 14:1 and ratios over 22:1 are also common.

Thermal Efficiency for Diesel Cycle

In general the thermal efficiency, η_{th}, of any heat engine is defined as the ratio of the work it does, W, to the heat input at the high temperature, Q_H.

$$\eta_{th} = \frac{W}{Q_H}$$

The thermal efficiency, η_{th}, represents the fraction of heat, Q_H, that is converted to work. Since energy is conserved according to the first law of thermodynamics and energy cannot be be converted to work completely, the heat input, Q_H, must equal the work done, W, plus the heat that must be dissipated as waste heat Q_C into the environment. Therefore we can rewrite the formula for thermal efficiency as:

$$\eta_{th} = \frac{W}{Q_H} = \frac{Q_H - Q_C}{Q_H} = 1 - \frac{Q_C}{Q_H}$$

The heat absorbed occurs during combustion of fuel-air mixture, when the spark occurs, roughly at constant volume. Since during an isochoric process there is no work done by or on the system, the first law of thermodynamics dictates $\Delta U = \Delta Q$. Therefore the heat added and rejected are given by:

$$Q_{add} = mc_p (T_3 - T_2)$$

$$Q_{out} = mc_v (T_4 - T_1)$$

Substituting these expressions for the heat added and rejected in the expression for thermal efficiency yields:

$$\eta_{th} = 1 - \frac{1 T_4 - T_1}{\kappa T_3 - T_2}$$

This equation can be rearranged to the form with the compression ratio and the cut-off ratio:

$$\eta_{Diesel} = 1 - \frac{1}{CR^{\kappa-1}} \left(\frac{\alpha^{\kappa} - 1}{\kappa(\alpha - 1} \right)$$

where:

- ηDiesel is the maximum thermal efficiency of a Diesel cycle.
- α is the cut-off ration V_3/V_2 (i.e. the ratio of volumes at the end and start of the combustion phase).
- CR is the compression ratio.
- κ = cp/cv = 1.4.

It is very useful conclusion, because it is desirable to achieve a high compression ratio to extract more mechanical energy from a given mass of the fuel. The air-standard Otto cycle thermal efficiency is also a function of compression ratio and κ.

$$\eta_{th} = 1 - \frac{T_4 - T_1}{T_3 - T_2} \rightarrow \eta_{Otto} = 1 - \left(\frac{V_2}{V_1} \right)^{\kappa-1} = 1 - \frac{1}{CR^{\kappa-1}}$$

When we compare these to formulae, it can be seen that for a given compression ratio (CR), the Otto cycle will be more efficient than the Diesel cycle. But diesel engines are usually be more efficient since they are able to operate at higher compression ratios.

In ordinary Otto engines the compression ratio has its limits. The compression ratio in a gasoline-powered engine will usually not be much higher than 10:1. Higher compression ratios will make gasoline engines subject to engine knocking, caused by autoignition an unburned mixture, if lower octane-rated fuel is used. In diesel engines there is minimal risk of autoignition of the fuel, because diesel engines are compression-ignition engines and there is no fuel in the cylinder at the beginning of the compression stroke.

Mean Effective Pressure – MEP

A parameter used by engineers to describe the performance of reciprocating piston engines is known as the mean effective pressure, or MEP. MEP is a very useful measure of an engine's capacity to do work that is independent of engine displacement. There are several types of MEP. These MEPs are defined by the location measurement and method of calculation (e.g. BMEP or IMEP).

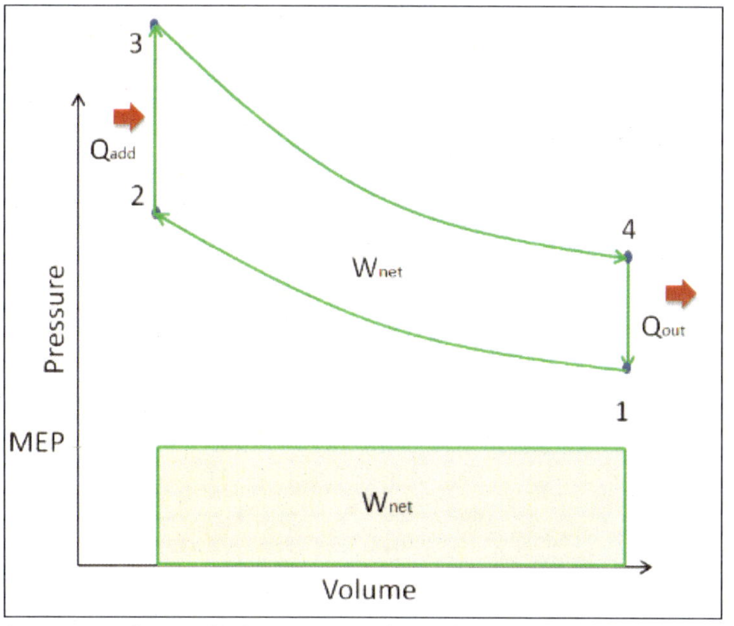

MEP is a very useful measure of an engine's capacity to do work that is independent of engine displacement.

In general, the mean effective pressure is the theoretical constant pressure that, if it acted on the piston during the power stroke, would produce the same net work as actually developed in one complete cycle. The MEP can be defined as:

$$MEP = \frac{net\ work\ for\ one\ cycle}{displacement\ volume}$$

For example, the net indicated mean effective pressure, known as $IMEP_n$ is equal to the mean effective pressure calculated from in-cylinder pressure (there must be this measurement) over the complete engine cycle. Note that, it is 720° for a four-stroke engine and 360° for a two-stroke engine.

Some examples:

- MEP of an atmospheric gasoline engine can range from 8 to 11 bar in the region of maximum torque.
- MEP of a turbocharged gasoline engine can range from 12 to 17 bar.
- MEP of an atmospheric diesel engine can range from 7 to 9 bar.
- MEP of a turbocharged diesel engine can range from 14 to 18 bar

For example, a four-stroke gasoline engine producing 200 N·m from 2 litres of displacement has a MEP of $(4\pi)(200\ N\cdot m)/(0.002\ m^3) = 1256000\ Pa = 12\ bar$. As can be seen, the MEP is useful characteristics of an engine. For two engines of equal displacement volume, the one with a higher MEP would produce the greater net work and, if the engines run at the same speed, greater power.

Diesel Cycle – Problem with Solution

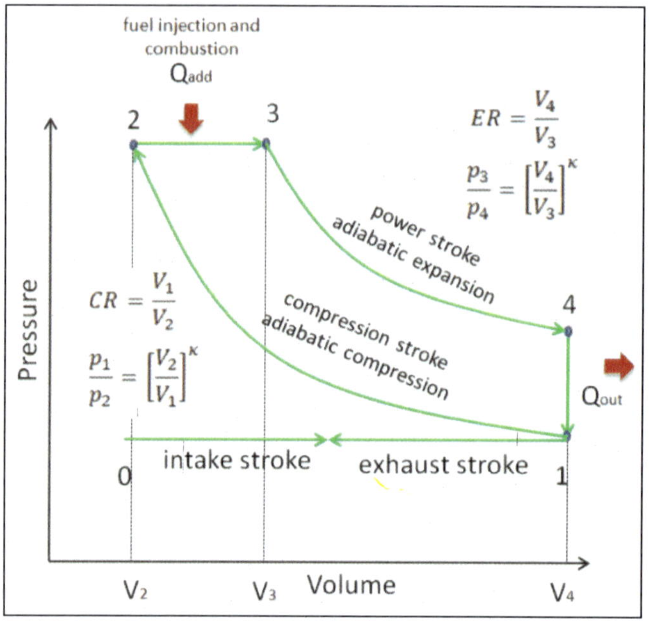

pV diagram of an ideal Diesel cycle.

Let assume the Diesel cycle, which is the one of most common thermodynamic cycles that can be found in automobile engines. One of key parameters of such engines is the change in volumes between top dead center (TDC) to bottom dead center (BDC). The ratio of these volumes (V_1 / V_2) is known as the compression ratio. Also the cut-off ratio V_3/V_2, which is the ratio of volumes at the end and start of the combustion phase.

In this example let assume the Diesel cycle with compression ratio of CR = 20 : 1 and cut-off ratio α = 2. The air is at 100 kPa = 1 bar, 20 °C (293 K), and the volume of the chamber is 500 cm³ prior to the compression stroke.

- Specific heat capacity at constant pressure of air at atmospheric pressure and room temperature: cp = 1.01 kJ/kgK.
- Specific heat capacity at constant volume of air at atmospheric pressure and room temperature: cv = 0.718 kJ/kgK.
- κ = cp/cv = 1.4.

Calculate:

- The mass of intake air.
- The temperature T2.
- The pressure p2.
- The temperature T3.
- The amount of heat added by burning of fuel-air mixture.
- The thermal efficiency of this cycle.
- The MEP.

Solution:

- At the beginning of calculations we have to determine the amount of gas in the cylinder before the compression stroke. Using the ideal gas law, we can find the mass:

  $pV = mR_{specific}T$

where,

p is the absolute pressure of the gas.

m is the mass of substance.

T is the absolute temperature.

V is the volume

Rspecific is the specific gas constant, equal to the universal gas constant divided by the molar mass (M) of the gas or mixture. For dry air Rspecific = 287.1 J.kg-1.K-1.

Therefore,

m = p1V1/RspecificT1 = (100000 × 500×10-6)/(287.1 × 293) = 5.95×10-4 kg

- In this problem all volumes are known:
  - $V_1 = V_4 = V_{max} = 500\times10^{-6}$ m^3 (0.5l)
  - $V_2 = V_{min} = V_{max} / CR = 25 \times10^{-6}$ m^3

  $(V_{max} - V_{min})$ x number of cylinders = total engine displacement.

Since the process is adiabatic, we can use the following p, V, T relation for adiabatic processes:

$$\left[\frac{V_2}{V_1}\right]^{\kappa} = \left[\frac{T_1}{T_2}\right]^{\frac{\kappa}{\kappa-1}}$$

$$\left[\frac{V_4}{V_3}\right]^{\kappa} = \left[\frac{T_3}{T_4}\right]^{\frac{\kappa}{\kappa-1}}$$

Thus,

$T_2 = T_1 . CR^{\kappa-1} = 293 . 20^{0.4} = 971$ K

Again, we can use the ideal gas law to find the pressure at the end of the compression stroke as:

$p_2 = mR_{specific}T_2 / V_2 = 5.95\times10^{-4} \text{ x } 287.1 \text{ x } 971 / 25 \times10^{-6} = 6635000$ Pa = 66.35 bar

Since process 2 → 3 occurs at constant pressure, the ideal gas equation of state gives.

$T_3 = (V_3/V_2) \text{ x } T_2 = 1942$ K

To calculate the amount of heat added by burning of fuel-air mixture, Q_{add}, we have to use the first law of thermodynamics for isobaric process, which states:

$Q_{add} = mc_p (T_3 - T_2) = 5.95\times10^{-4}$ x 1010 x 971 = 583.5 J

Thermal efficiency for this Diesel cycle:

The thermal efficiency of Diesel cycle is a function of the compression ratio, the cut-off ratio and κ:

$$\eta_{Diesel} = 1 - \frac{1}{CR^{\kappa-1}}\left(\frac{\alpha^{\kappa} - 1}{\kappa(\alpha - 1}\right)$$

Where,

η_{Diesel} is the maximum thermal efficiency of a Diesel cycle.

α is the cut-off ration V_3/V_2 (i.e. the ratio of volumes at the end and start of the combustion phase).

CR is the compression ratio.

$\kappa = cp/cv = 1.4$.

For this example:

$\eta_{Diesel} = 0.6467 = 64.7\%$

The MEP was defined as:

$$MEP = \frac{net\ work\ for\ one\ cycle}{displacement\ volume}$$

It this equation the displacement volume is equal to $V_{max} - V_{min}$. The net work for one cycle can be calculated using the heat added and the thermal efficiency:

$W_{net} = Q_{add} \cdot \eta_{Otto} = 583.5 \text{ x } 0.6467 = 377.3$ J

MEP = 377.3 / ($500\times10^{-6} - 25 \times10^{-6}$) = 794.3 kPa = 7.943 bar.

Applications

CHAPTER 5

There are a wide-range of applications of gas turbines in electric power generation in industries as well as in marine, locomotive, and automotive propulsion. This chapter closely examines these applications of gas turbines to provide an extensive understanding of the subject.

ENGINE FOR SURFACE TRANSPORTATION

Gas turbines are often used on ships, locomotives, helicopters, tanks, and to a lesser extent, on cars, buses, and motorcycles.

A key advantage of jets and turboprops for airplane propulsion - their superior performance at high altitude compared to piston engines, particularly naturally aspirated ones - is irrelevant in most automobile applications. Their power-to-weight advantage, though less critical than for aircraft, is still important.

Gas turbines offer a high-powered engine in a very small and light package. However, they are not as responsive and efficient as small piston engines over the wide range of RPMs and powers needed in vehicle applications. In series hybrid vehicles, as the driving electric motors are mechanically detached from the electricity generating engine, the responsiveness, poor performance at low speed and low efficiency at low output problems are much less important. The turbine can be run at optimum speed for its power output, and batteries and ultracapacitors can supply power as needed, with the engine cycled on and off to run it only at high efficiency. The emergence of the continuously variable transmission may also alleviate the responsiveness problem.

Turbines have historically been more expensive to produce than piston engines, though this is partly because piston engines have been mass-produced in huge quantities for decades, while small gas turbine engines are rarities; however, turbines are mass-produced in the closely related form of the turbocharger.

The turbocharger is basically a compact and simple free shaft radial gas turbine which is driven by the piston engine's exhaust gas. The centripetal turbine wheel drives a centrifugal compressor wheel through a common rotating shaft. This wheel supercharges the engine air intake to a degree that can be controlled by means of a wastegate or by dynamically modifying the turbine housing's geometry (as in a VGT turbocharger). It mainly serves as a power recovery device which converts a great deal of otherwise wasted thermal and kinetic energy into engine boost.

Turbo-compound engines (actually employed on some trucks) are fitted with blow down turbines which are similar in design and appearance to a turbocharger except for the turbine shaft being mechanically or hydraulically connected to the engine's crankshaft instead of to a centrifugal compressor, thus providing additional power instead of boost. While the turbocharger is a pressure turbine, a power recovery turbine is a velocity one.

MAZ-7907, a Transporter erector launcher with a Turbine-electric transmission.

Passenger Road Vehicles (Cars, Bikes, and Buses)

A number of experiments have been conducted with gas turbine powered automobiles, the largest by Chrysler.More recently, there has been some interest in the use of turbine engines for hybrid electric cars. For instance, a consortium led by micro gas turbine company Bladon Jets has secured investment from the Technology Strategy Board to develop an Ultra Lightweight Range Extender (ULRE) for next-generation electric vehicles. The objective of the consortium, which includes luxury car maker Jaguar Land Rover and leading electrical machine company SR Drives, is to produce the world's first commercially viable - and environmentally friendly - gas turbine generator designed specifically for automotive applications.

The common turbocharger for gasoline or diesel engines is also a turbine derivative.

Concept Cars

The 1950 Rover JET1.

The first serious investigation of using a gas turbine in cars was in 1946 when two engineers, Robert Kafka and Robert Engerstein of Carney Associates, a New York engineering firm, came up with the concept where a unique compact turbine engine design would provide power for a rear wheel drive car.

In 1950, designer F.R. Bell and Chief Engineer Maurice Wilks from British car manufacturers Rover unveiled the first car powered with a gas turbine engine. The two-seater JET1 had the engine positioned behind the seats, air intake grilles on either side of the car, and exhaust outlets on the top of the tail. During tests, the car reached top speeds of 140 km/h (87 mph), at a turbine speed of 50,000 rpm. The car ran on petrol, paraffin (kerosene) or diesel oil, but fuel consumption problems proved insurmountable for a production car. It is on display at the London Science Museum.

A French turbine powered car, the Socema-Gregoire, was displayed at the October 1952 Paris Auto Show. It was designed by the French engineer Jean-Albert Grégoire.

GM Firebird I.

The first turbine-powered car built in the US was the GM Firebird I which began evaluations in 1953. While photos of the Firebird I may suggest that the jet turbine's thrust propelled the car like an aircraft, the turbine actually drove the rear wheels. The Firebird 1 was never meant as a commercial passenger car and was solely built for testing & evaluation as well as public relation purposes.

Engine compartment of a Chrysler 1963 Turbine car.

Starting in 1954 with a modified Plymouth, the American car manufacturer Chrysler demonstrated several prototype gas turbine-powered cars from the early 1950s through the early 1980s. Chrysler built fifty Chrysler Turbine Cars in 1963 and conducted the only consumer trial of gas turbine-powered cars. Each of their turbines employed a unique rotating recuperator, referred to as a regenerator that increased efficiency.

In 1954 FIAT unveiled a concept car with a turbine engine, called Fiat Turbina. This vehicle, looking like an aircraft with wheels, used a unique combination of both jet thrust and the engine driving the wheels. Speeds of 282 km/h (175 mph) were claimed.

The original General Motors Firebird was a series of concept cars developed for the 1953, 1956 and 1959 Motorama auto shows, powered by gas turbines.

As a result of the U.S. Clean Air Act Amendments of 1970, research was funded into developing automotive gas turbine technology. Design concepts and vehicles were conducted by Chrysler, General Motors, Ford (in collaboration with AiResearch), and American Motors (in conjunction with Williams Research). Long-term tests were conducted to evaluate comparable cost efficiency. Several AMC Hornets were powered by a small Williams regenerative gas turbine weighing 250 lb (113 kg) and producing 80 hp (60 kW; 81 PS) at 4450 rpm.

Toyota demonstrated several gas turbine powered concept cars, such as the Century gas turbine hybrid in 1975, the Sports 800 Gas Turbine Hybrid in 1979 and the GTV in 1985. No production vehicles were made. The GT24 engine was exhibited in 1977 without a vehicle.

In the early 1990s, Volvo introduced the Volvo Environmental Concept Car(ECC) which was a gas turbine powered hybrid car.

In 1993 General Motors introduced the first commercial gas turbine powered hybrid vehicle—as a limited production run of the EV-1 series hybrid. A Williams International 40 kW turbine drove an alternator which powered the battery-electric powertrain. The turbine design included a recuperator. In 2006, GM went into the EcoJet concept car project with Jay Leno.

At the 2010 Paris Motor Show Jaguar demonstrated its Jaguar C-X75 concept car. This electrically powered supercar has a top speed of 204 mph (328 km/h) and can go from 0 to 62 mph (0 to 100 km/h) in 3.4 seconds. It uses Lithium-ion batteries to power four electric motors which combine to produce 780 bhp. It will travel 68 miles (109 km) on a single charge of the batteries, and uses a pair of Bladon Micro Gas Turbines to recharge the batteries extending the range to 560 miles (900 km).

Racing Cars

The 1967 *STP Oil Treatment Special* on display at the Indianapolis Motor Speedway Hall of Fame Museum, with the Pratt & Whitney gas turbine shown.

The first race car (in concept only) fitted with a turbine was in 1955 by a US Air Force group as a hobby project with a turbine loaned them by Boeing and a race car owned by Firestone Tire & Rubber company. The first race car fitted with a turbine for the goal of actual racing was by Rover and the BRM Formula One team joined forces to produce the Rover-BRM, a gas turbine powered coupe, which entered the 1963 24 Hours of Le Mans, driven by Graham Hill and Richie Ginther. It averaged 107.8 mph (173.5 km/h) and had a top speed of 142 mph (229 km/h). American Ray Heppenstall joined Howmet Corporation and McKee Engineering together to develop their own gas turbine sports car in 1968, the Howmet TX, which ran several American and European events, including two wins, and also participated in the 1968 24 Hours of Le Mans. The cars used Continental gas turbines, which eventually set six FIA land speed records for turbine-powered cars.

For open wheel racing, 1967's revolutionary STP-Paxton Turbocar fielded by racing and entrepreneurial legend Andy Granatelli and driven by Parnelli Jones nearly won the Indianapolis 500; the Pratt & Whitney ST6B-62 powered turbine car was almost a lap ahead of the second place car when a gearbox bearing failed just three laps from the finish line.

A 1968 Howmet TX, the only turbine-powered race car to have won a race.

Buses

The arrival of the Capstone Microturbine has led to several hybrid bus designs, starting with HEV-1 by AVS of Chattanooga, Tennessee in 1999, and closely followed by Ebus and ISE Research in California, and DesignLine Corporation in New Zealand (and later the United States). AVS turbine hybrids were plagued with reliability and quality control problems, resulting in liquidation of AVS in 2003. The most successful design by Designline is now operated in 5 cities in 6 countries, with over 30 buses in operation worldwide, and order for several hundred being delivered to Baltimore, and New York City.

Brescia Italy is using serial hybrid buses powered by microturbines on routes through the historical sections of the city.

Motorcycles

The MTT Turbine Superbike appeared in 2000 (hence the designation of Y2K Superbike by MTT) and is the first production motorcycle powered by a turbine engine - specifically, a Rolls-Royce Allison model 250 turboshaft engine, producing about 283 kW (380 bhp). Speed-tested to 365 km/h or 227 mph (according to some stories, the testing team ran out of road during the test), it holds the Guinness World Record for most powerful production motorcycle and most expensive production motorcycle, with a price tag of US$185,000.

Trains

Several locomotive classes have been powered by gas turbines, the most recent incarnation being Bombardier's JetTrain.

Tanks

Marines from 1st Tank Battalion load a Honeywell AGT1500 multi-fuel turbine back into an M1 Abrams tank at Camp Coyote, Kuwait.

The Third Reich *Wehrmacht Heer's* development division, the Heereswaffenamt (Army Ordnance Board), studied a number of gas turbine engine designs for use in tanks starting in mid-1944. The first gas turbine engine design intended for use in armored fighting vehicle propulsion, the BMW 003-based GT 101, was meant for installation in the Panther tank.

The second use of a gas turbine in an armored fighting vehicle was in 1954 when a unit, PU2979, specifically developed for tanks by C. A. Parsons & Co., was installed and trialled in a British Conqueror tank. The Stridsvagn 103 was developed in the 1950s and was the first mass-produced main battle tank to use a turbine engine. Since then, gas turbine engines have been used as APUs in some tanks and as main powerplants in Soviet/Russian T-80s and U.S. M1 Abrams tanks, among others. They are lighter and smaller than diesels at the same sustained power output but the models installed to date are less fuel efficient than the equivalent diesel, especially at idle, requiring more fuel to achieve the same combat range. Successive models of M1 have addressed this problem with battery packs or secondary generators to power the tank's systems while stationary, saving fuel by reducing the need to idle the main turbine. T-80s can mount three large external fuel drums to extend their range. Russia has stopped production of the T-80 in favor of the diesel-powered T-90 (based on the T-72), while Ukraine has developed the diesel-powered T-80UD and T-84 with nearly the power of the gas-turbine tank. The French Leclerc MBT's diesel powerplant features the "Hyperbar" hybrid supercharging system, where the engine's turbocharger is completely replaced with a small gas turbine which also works as an assisted diesel exhaust turbocharger, enabling engine RPM-independent boost level control and a higher peak boost pressure to be reached (than with ordinary turbochargers). This system allows a smaller displacement and lighter engine to be used as the tank's powerplant and effectively removes turbo lag. This special gas turbine/turbocharger can also work independently from the main engine as an ordinary APU.

A turbine is theoretically more reliable and easier to maintain than a piston engine since it has a simpler construction with fewer moving parts, but in practice, turbine parts experience a higher wear rate due to their higher working speeds. The turbine blades are highly sensitive to dust and fine sand so that in desert operations air filters have to be fitted and changed several times daily. An improperly fitted filter, or a bullet or shell fragment that punctures the filter, can damage the engine. Piston engines (especially if turbocharged) also need well-maintained filters, but they are more resilient if the filter does fail.

Like most modern diesel engines used in tanks, gas turbines are usually multi-fuel engines.

The main demands for the transport gas turbine units were determined. Road transport and ship gas turbine engines development periods; the most rational fields of their application, their advantages and disadvantages in comparison with other engine types were discussed. Thermodynamic cycle analyses and principle schemes of gas turbine units were carried out. The construction examples of most typical gas turbine engines in different classes of ship, wheel and truck machines and railway transport power units were given. The main problems under these engine operation and their probable solution ways were shown. The future transport gas turbine development was discussed.

Presently available gas turbines (GT) find an ever-growing use in marine and land vehicle power plants due to many advantages they can offer over other thermal engines. The principal advantages of GT are: best specific mass and overall dimensions, abandonment of the liquid-cooling systems, simple maintenance, good cold starting characteristics, clean exhaust due to high air-fuel ratio etc.. Some advantages of GT are of critical importance in the selection of this engine for a certain type of vehicle. Thus, carrying capacity of some heavy dumpers may exceed 100 t. GT engines having output of 1000 kW and more are available and up to 1500 kW in the nearest future. Diesels of similar output level have excessive weight and overall dimensional data, worse torque and exhaust with toxic characteristics, more time-consuming maintenance than that for GT. The most important requirements for power plants of military vehicles are: space saving, good cold starting characteristics, simplicity of the operational control. Automotive engines must release minimum of the toxic components in exhaust gases and have simple maintenance. GT application to the heavy duty automotives and trucks having mass 25 – 50 t must have 350 – 400 kW power unit and it will promote driving, simplify maintenance, permit abandonment of the liquid-cooling systems, improve reliability of the power-unit.

Advantages stipulate application of GT in marine and traffic power plants. Turbine engines are used in ships, for which little specific weight and overall dimensions have essential significance, possibilities for any engine parts immediate replacement maneuverability.

GT are extensively used in the navy and in dynamic support principal ships – hovercrafts and hydrofoils. GT have better specific data in comparison with steamturbine unit due to the elimination of the feed water heating boiler, less number of auxiliary instruments. GT demonstrates low cost of service in operation, ease of automation. In comparison with reciprocating internal combustion engines marine GT is also profitable in size and weight, may use cheaper fuel (oil fuel for instance), is simpler in service.

However marine and transport GT are superior to diesel power plants in fuel economy when applied to the ships and heavy-duty automobiles. It also relates to the steam turbine marine power plants and gas internal combustion engines, which are the main type of low power engines for land transport.

Requirements of Transport Power Plants

The term "power plant" must be understood as a combination of engines with systems providing its operation in different conditions of operation. Those combinations include: fuel and lubrication systems, starting device control and testing system, air delivery and cleaning system, cooling system, exhaust gas devices system etc. Besides power generation for propulsion, as a general purpose, additional requirements imposed on the power plant are provisions with electricity, steam and compressed air for service aid. These requirements are achieved by the use of special-purpose equipment (electric generator, air compressor) driven by the shaft of the main engine or by the auxiliary power unit (APU). Application of the turbine or piston APU raises the overhaul period of the main engine and improves economy of the vehicle in operation. It can be achieved by disengaging of the main engine or by reducing its operation period when the ship or vehicle is inoperative. APU is also used to facilitate main engine starting.

Power plants may also contain additional devices providing characteristics of supplementary modes for instance, for retardation or reverse (for ships). Engine purpose must be taken into account while developing the base engine models for transport or their modifications. These requirements depend principally on the load and operation modes. Mutual influence of the engine and environment must also be taken into account.

Requirements Determined by Load Distinctions

The main engine of transport is connected with the propelling agent through a reduction gear which is used to match speeds of the engine shaft with that of the propelling shaft. Electrical transmission (having generator and motor) or hydraulic drives are used less often. Transmission must afford reverse of transport or possibility of a variety gear ratio (in land vehicles). Marine power plant transmission can be used for power summation of several engines or for power distribution to different propelling agents.

Fixed blade propeller characteristics are approximately proportionate to the squared torque at the given speed. Therefore, torque sharply decreases with propeller speed drop.

By contrast, characteristics of the load at an automotive engine shaft or at an adjustable propeller present a field of possible modes. Thus different amounts of automotive engine torque can correspond to each speed of the carrying wheel depending on ground characteristics, road slope, and other factors.

At straight-line traffic of the vehicle having mass G and speed V on the road with level or slope angle α, the torque at the carrying wheel is:

- Overcoming slope force $P1 = G \cdot \sin\alpha$, where angle α depends on the transport purpose,
- Overcoming ground resistant force $P2 = G \cdot f_{gr} \cdot \cos\alpha$, where f_{gr} is the ground resistance coefficient (approximately 0.02 for asphalt, 0.03...0.05 for dry soil road, 0.08...0.1 for sand road),
- Aerodynamic resistance force $P3 = k \cdot F \cdot V_2$, where F – automotive cross section,
- k – air resistance coefficient, depending on the shape and surface quality.

Taking into consideration $f_r = f_{gr} \cdot \cos\alpha + \sin\alpha$, where f_r - resistance coefficient for the automobile steady-state traffic, carrying wheel torque is $M_{cw} = R_{cw} \bullet (G \bullet f_r + k \cdot F \cdot V^2)$, where R_{cw} – radius of the carrying wheel. Torque at the aft shaft $Ma = Mcw \bullet (i_{tr} \cdot \eta_{tr} \cdot \eta_{ch})$, where $i\,t_r$ – transmission gear ratio; η_{tr}, η_{ch} – transmission and chassis efficiency.

Wide load range for automobile include: starting, level and slope traffic in different road conditions. It becomes necessary to use the gearbox, i.e. increment transmission application dependents on the automobile traffic speed. This requirement cannot be satisfied by the application of a single shaft GT, combined with constant gear ratio transmission whose characteristics relate to 60 – 100 % of the aft shaft speed range (at n < 60% the inadmissible rise of gas temperature begins). It also leads to the rapid drop of torque with the reduction of rotation speed.

Two or three-shaft GT with free power turbine (load turbine) in this respect have acceptable propulsion torque characteristics in full speed range. Practically, it rises proportionally to the drop of the shaft speed, up to shut-down without engine off.

In project stage of the GT transient mode, conditions must be taken into consideration when necessary to provide acceleration or deceleration, start or stop of the engine. These modes are of first-rate importance, because transient modes in operation often define the dynamics of the vehicle concerned with the rate of acceleration or deceleration. Important and evident requirement of an engine is maneuverability of vehicle or ship, including propulsion stop and reverse. Introducing special devices can attain retarding modes of the transport power plants (for instance: retarder in automobile, reverse turbine in ship power plant). The other way is to apply special devices such as variable nozzle and shaft lock-up clutch. Note should be taken that GT provides greater moment of inertia to the shaft than piston engine in spite of the lower moment of inertia of the power turbine wheel. Therefore the equivalent retardation dynamics of the vehicle can be attained when GT power plant at similar moment can afford as twice torque as power plant with diesel.

Requirements Determined by Operational Conditions

The main operational requirements of transport power units are as follows: economical operation, reliability, long service live before the overhaul, simplicity of service and low cost of operation. The dimensional value and the mass of a power installation as well as the possibility of using various types of fuels (including cheap ones), the oil consumption quantity and the necessity of utilizing other special fluids, the possibility of putting the engine into operation easily under different conditions, including negative air temperature, the simplicity of controlling and other factors are of great importance nowadays.

The engine with its systems has to be located under design limitations attributed to its application in the particular vehicle. It should have a good access to places of maintenance and regulations as well (fuel instrumentation, filters).

The run distance of transport vehicle and the floating of the vessel are defined by the quantity of fuel, by the unit power and the specific fuel consumption of an engine (or by effective efficiency). The efficiency is of primary importance while defining the operational traveling fuel consumption, which in its turn depends on both the profitability of the engine under different operation working conditions and on the time span of these conditions. Depending on the type and purpose of the vehicle (vessel) and on its operation conditions, the engine operational schedules are essentially different.

For example, a city bus on a line or a car in the urban cycle, a tanker on a long distance line or a boat at coast guard appear to have absolutely different operating conditions and time span for engine load operation. The traveling fuel consumption is greatly influenced by the idle engine duration and by the idle fuel consumption.

The engine has to meet the requirements of the customers in its reliability. This integrated concept is composed of both long service life and the original life before the capital repair, of a non-stop as the accessible number of failures during a certain interval of operational time. It should also include the concept of maintainability, the ability of control and keeping quality.

The engine design should be calculated for the effect of transport loading, rolls and trims taking place while the vehicle is traveling. The air feeding system should prevent water or abrasive particles (dust) to get into the engine in order to avoid erosion or contamination of the compressor and the turbine blades.

In order to decrease the engine designing and production cost the possibility of their multipurpose application as aviation, transport and ship engines are often discussed.

Gas Turbine Locomotive

A locomotive that has a gas-turbine engine or an engine combining a free-piston gas generator and a gas turbine. Nearly all existing gas-turbine locomotives have a single-shaft gas-turbine unit with an open cycle and with electrical drive.

Gas-turbine locomotives first appeared in the USA in 1948. In 1969 the Union Pacific Railroad had as many as 50 gasturbine locomotives in use, with engines with a power of 3,300 kilowatts (kW), or 4,500 hp, and 6,300 kW (8,500 hp). Gas-turbine locomotives have also been built in Great Britain, Sweden, Switzerland, and Czechoslovakia. The first such locomotives in the Soviet Union were put into use in 1965.

The power plant of a gas-turbine locomotive with electrical drive consists of a gas turbine, a compressor, a DC generator, and traction motors. The generator supplies electrical energy to the traction motors, which are usually located on each drive axle of the locomotive.

The type of gas-turbine engine and its intended application determine the mode of power transmission from the turbine shaft to the drive wheels. If the engine has one shaft, an electrical drive like that in a diesel locomotive is used. So-called rigid drive, which may be used in gas-turbine locomotives with multishaft engines, may be mechanical (main reduction gear, universal joints, and axial reduction gears) or AC electrical (synchronous generators and asynchronous squirrel-cage motors). Gas-turbine locomotives usually have a diesel starter with a power of 150-240 kW (200-300 hp), whose main function is to raise the speed of rotation of the generator to the point at which the compressor begins to supply air to the combustion chamber. The starter also propels the locomotive when it is running without its train; and it supplies power to certain auxiliary equipment. The gasturbine unit on a locomotive usually operates on heavy liquid fuel or gas.

Gas-turbine locomotives have a number of advantages not only in comparison with steam locomotives but also, in several respects, in comparison with diesel locomotives. For example, the specific weight of a gas-turbine locomotive that is, the weight per unit of power is about 50 percent of the weight for a steam locomotive and about 75 percent of that for a diesel locomotive. The compactness of a gasturbine unit makes it possible to reduce the length of the locomotive to about onehalf that of a diesel locomotive of equal power. The power plant of the gas-turbine locomotive does not require water. The simplicity of construction of a gas-turbine system ensures its reliability and uninterrupted operation; it also facilitates maintenance and routine repairs. The operation of a gas-turbine locomotive is simply the regulation of the supply of fuel to the combustion chamber. The proper operation of the electrical drive is maintained automatically.

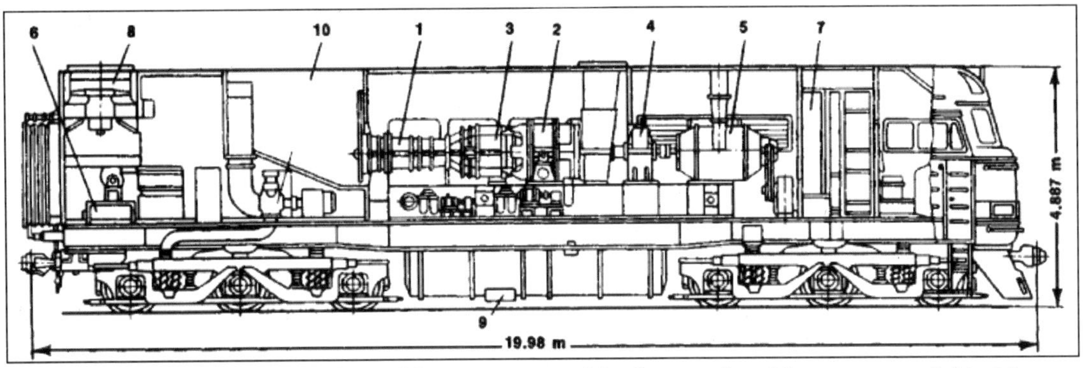

Configuration of the power plant of the G1-01 gas-turbine locomotive: (1) compressor, (2) turbine, (3) combustion chamber, (4) reduction gear, (5) main generators, (6) auxiliary diesel engine, (7) high-voltage chambers, (8) cooler, (9) fuel tank, (10) braking compressor.

Gas-turbine locomotives produced at the Kolomna Locomotive Plant (model G1-O1; figure) are noted for the high reliability of their engines, for simplicity of maintenance and repair, and for their ability to operate on heavy fuels.

Gas-turbine locomotive engineering still has not left the experimental stage. Gas-turbine locomotives are still comparatively inefficient (about half as efficient as diesel locomotives); attempts are being made to increase their efficiency. The Lugansk Locomotive Plant has built an experimental gas-turbine locomotive with a free-piston gas generator and a power of 2,200 kW (3,000 hp). Its power plant consists of four electrical generators working from one gas turbine, a hydromechanical drive, and auxiliary equipment. Similar experimental gas-turbine locomotives have also been built in France and Sweden; their efficiency is as high as 30-32 percent.

APPLICATION IN ELECTRICITY GENERATION

In electricity generating applications the turbine is used to drive a synchronous generator which provides the electrical power output but because the turbine normally operates at very high rotational speeds of 12,000 r.p.m or more it must be connected to the generator through a high ratio reduction gear since the generators run at speeds of 1,000 or 1,200 r.p.m. depending on the AC frequency of the electricity grid.

Turbine Configurations

Gas turbine power generators are used in two basic configurations:

- Simple Systems consisting of the gas turbine driving an electrical power generator.

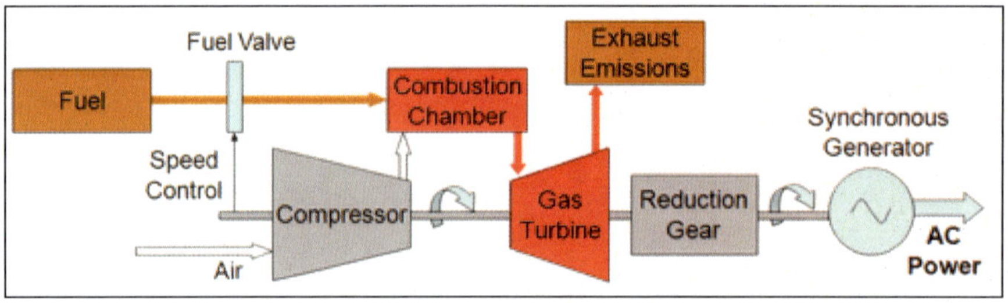

Gas Turbine Electric Power Generation.

- Combined Cycle Systems which are designed for maximum efficiency in which the hot exhaust gases from the gas turbine are used to raise steam to power a steam turbine with both turbines being connected to electricity generators.

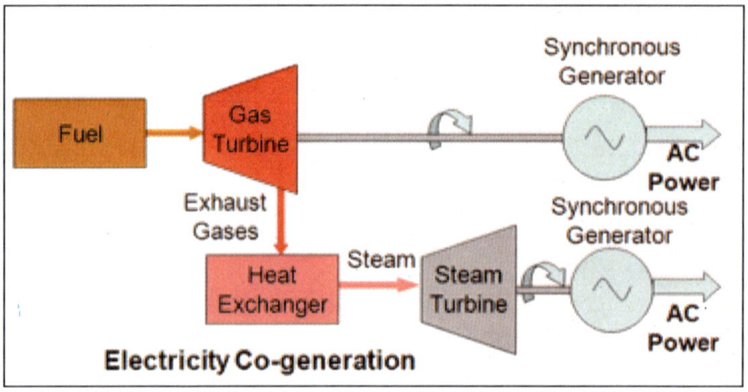

Electricity Co-generation.

Turbine Performance

Turbine Power Output

To minimise the size and weight of the turbine for a given output power, the output per pound of airflow should be maximised. This is obtained by maximising the air flow through the turbine which in turn depends on maximising the pressure ratio between the air inlet and exhaust outlet. The main factor governing this is the pressure ratio across the compressor which can be as high as 40:1 in modern gas turbines. In simple cycle applications, pressure ratio increases translate into efficiency gains at a given firing temperature, but there is a limit since increasing the pressure ratio means that more energy will be consumed by the compressor.

System Efficiency

Thermal efficiency is important because it directly affects the fuel consumption and operating costs.

Simple Cycle Turbines

A gas turbine consumes considerable amounts of power just to drive its compressor. As with all cyclic heat engines, a higher maximum working temperature in the machine means greater efficiency (Carnot's Law), but in a turbine it also means that more energy is lost as waste heat through the hot exhaust gases whose temperatures are typically well over 1,000 °C. Consequently simple cycle turbine efficiencies are quite low. For heavy plant, design efficiencies range between 30% and 40%. (The efficiencies of aero engines are in the range 38% and 42% while low power microturbines (<100kW) achieve only 18% to 22%). Although increasing the firing temperature increases the output power at a given pressure ratio, there is also a sacrifice of efficiency due to the increase in losses due to the cooling air required to maintain the turbine components at reasonable working temperatures.

Combined Cycle Turbines

It is however possible to recover energy from the waste heat of simple cycle systems by using the exhaust gases in a hybrid system to raise steam to drive a steam turbine electricity generating set. In such cases the exhaust temperature may be reduced to as low as 140 °C enabling efficiencies of up to 60% to be achieved in combined cycle systems.

In combined-cycle applications, pressure ratio increases have a less pronounced effect on the efficiency since most of the improvement comes from increases in the Carnot thermal efficiency resulting from increases in the firing temperature.

Thus simple cycle efficiency is achieved with high pressure ratios. Combined cycle efficiency is obtained with more modest pressure ratios and greater firing temperatures.

Fuels

One further advantage of gas turbines is their fuel flexibility. They can be adapted to use almost any flammable gas or light distillate petroleum products such as gasoline (petrol), diesel and kerosene (paraffin) which happen to be available locally, though natural gas is the most commonly used fuel. Crude and other heavy oils and can also be used to fuel gas turbines if they are first heated to reduce their viscosity to a level suitable for burning in the turbine combustion chambers.

Applications

Gas turbines can be used for large scale power generation. Examples are applications delivering 600 MW or more from a 400 MW gas turbine coupled to a 200 MW steam turbine in a co-generating installation. Such installations are not normally used for base load electricity generation, but for bringing power to remote sites such as oil and gas fields. They do however find use in the major electricity grids in peak shaving applications to provide emergency peak power.

Low power gas turbine generating sets with capacities up to 5 MW can be accommodated in transportation containers to provide mobile emergency electricity supplies which can delivered by truck to the point of need.

Environmental Issues

Almost all gas turbine installations use fossil fuels.

References

- Reverse gas turbine, 152 pp. S-Petersburg, "Sudostroenie" (In Russian)
- Gas-turbines: mpoweruk.com, Retrieved 28 August, 2019
- Marine Gas Turbines, 359 pp A Wiley – Interscience Publication John Wiley & Sons, N.Y.- London-Sydney – Toronto
- Gas-Turbine-Locomotive: encyclopedia2.thefreedictionary.com, Retrieved 14 May, 2019
- Kay, Antony L. (2002). German jet engine and gas turbine development 1930 - 1945. Airlife. ISBN 9781840372946

PERMISSIONS

All chapters in this book are published with permission under the Creative Commons Attribution Share Alike License or equivalent. Every chapter published in this book has been scrutinized by our experts. Their significance has been extensively debated. The topics covered herein carry significant information for a comprehensive understanding. They may even be implemented as practical applications or may be referred to as a beginning point for further studies.

We would like to thank the editorial team for lending their expertise to make the book truly unique. They have played a crucial role in the development of this book. Without their invaluable contributions this book wouldn't have been possible. They have made vital efforts to compile up to date information on the varied aspects of this subject to make this book a valuable addition to the collection of many professionals and students.

This book was conceptualized with the vision of imparting up-to-date and integrated information in this field. To ensure the same, a matchless editorial board was set up. Every individual on the board went through rigorous rounds of assessment to prove their worth. After which they invested a large part of their time researching and compiling the most relevant data for our readers.

The editorial board has been involved in producing this book since its inception. They have spent rigorous hours researching and exploring the diverse topics which have resulted in the successful publishing of this book. They have passed on their knowledge of decades through this book. To expedite this challenging task, the publisher supported the team at every step. A small team of assistant editors was also appointed to further simplify the editing procedure and attain best results for the readers.

Apart from the editorial board, the designing team has also invested a significant amount of their time in understanding the subject and creating the most relevant covers. They scrutinized every image to scout for the most suitable representation of the subject and create an appropriate cover for the book.

The publishing team has been an ardent support to the editorial, designing and production team. Their endless efforts to recruit the best for this project, has resulted in the accomplishment of this book. They are a veteran in the field of academics and their pool of knowledge is as vast as their experience in printing. Their expertise and guidance has proved useful at every step. Their uncompromising quality standards have made this book an exceptional effort. Their encouragement from time to time has been an inspiration for everyone.

The publisher and the editorial board hope that this book will prove to be a valuable piece of knowledge for students, practitioners and scholars across the globe.

INDEX